HPLC in the
Pharmaceutical Industry

DRUGS AND THE PHARMACEUTICAL SCIENCES

A Series of Textbooks and Monographs

Edited by
James Swarbrick
*School of Pharmacy
University of North Carolina
Chapel Hill, North Carolina*

Volume 1. PHARMACOKINETICS, *Milo Gibaldi and Donald Perrier* (out of print)

Volume 2. GOOD MANUFACTURING PRACTICES FOR PHARMACEUTICALS: A PLAN FOR TOTAL QUALITY CONTROL, *Sidney H. Willig, Murray M. Tuckerman, and William S. Hitchings IV* (out of print)

Volume 3. MICROENCAPSULATION, *edited by J. R. Nixon*

Volume 4. DRUG METABOLISM: CHEMICAL AND BIOCHEMICAL ASPECTS, *Bernard Testa and Peter Jenner*

Volume 5. NEW DRUGS: DISCOVERY AND DEVELOPMENT, *edited by Alan A. Rubin*

Volume 6. SUSTAINED AND CONTROLLED RELEASE DRUG DELIVERY SYSTEMS, *edited by Joseph R. Robinson*

Volume 7. MODERN PHARMACEUTICS, *edited by Gilbert S. Banker and Christopher T. Rhodes*

Volume 8. PRESCRIPTION DRUGS IN SHORT SUPPLY: CASE HISTORIES, *Michael A. Schwartz*

Volume 9. ACTIVATED CHARCOAL: ANTIDOTAL AND OTHER MEDICAL USES, *David O. Cooney*

Volume 10. CONCEPTS IN DRUG METABOLISM (in two parts), *edited by Peter Jenner and Bernard Testa*

Volume 11. PHARMACEUTICAL ANALYSIS: MODERN METHODS (in two parts), *edited by James W. Munson*

Volume 12. TECHNIQUES OF SOLUBILIZATION OF DRUGS, *edited by Samuel H. Yalkowsky*

Volume 13. ORPHAN DRUGS, *edited by Fred E. Karch*

Volume 14. NOVEL DRUG DELIVERY SYSTEMS: FUNDAMENTALS, DEVELOPMENTAL CONCEPTS, BIOMEDICAL ASSESSMENTS, *Yie W. Chien*

Volume 15. PHARMACOKINETICS, Second Edition, Revised and Expanded, *Milo Gibaldi and Donald Perrier*

Volume 16. GOOD MANUFACTURING PRACTICES FOR PHARMACEUTICALS: A PLAN FOR TOTAL QUALITY CONTROL, Second Edition, Revised and Expanded, *Sidney H. Willig, Murray M. Tuckerman, and William S. Hitchings IV*

Volume 17. FORMULATION OF VETERINARY DOSAGE FORMS, *edited by Jack Blodinger*

Volume 18. DERMATOLOGICAL FORMULATIONS: PERCUTANEOUS ABSORPTION, *Brian W. Barry*

Volume 19. THE CLINICAL RESEARCH PROCESS IN THE PHARMACEUTICAL INDUSTRY, *edited by Gary M. Matoren*

Volume 20. MICROENCAPSULATION AND RELATED DRUG PROCESSES, *Patrick B. Deasy*

Volume 21. DRUGS AND NUTRIENTS: THE INTERACTIVE EFFECTS, *edited by Daphne A. Roe and T. Colin Campbell*

Volume 22. BIOTECHNOLOGY OF INDUSTRIAL ANTIBIOTICS, *Erick J. Vandamme*

Volume 23. PHARMACEUTICAL PROCESS VALIDATION, *edited by Bernard T. Loftus and Robert A. Nash*

Volume 24. ANTICANCER AND INTERFERON AGENTS: SYNTHESIS AND PROPERTIES, *edited by Raphael M. Ottenbrite and George B. Butler*

Volume 25. PHARMACEUTICAL STATISTICS: PRACTICAL AND CLINICAL APPLICATIONS, *Sanford Bolton*

Volume 26. DRUG DYNAMICS FOR ANALYTICAL, CLINICAL, AND BIOLOGICAL CHEMISTS, *Benjamin J. Gudzinowicz, Burrows T. Younkin, Jr., and Michael J. Gudzinowicz*

Volume 27. MODERN ANALYSIS OF ANTIBIOTICS, *edited by Adorjan Aszalos*

Volume 28. SOLUBILITY AND RELATED PROPERTIES, *Kenneth C. James*

Volume 29. CONTROLLED DRUG DELIVERY: FUNDAMENTALS AND APPLICATIONS, Second Edition, Revised and Expanded, *edited by Joseph R. Robinson and Vincent H. L. Lee*

Volume 30. NEW DRUG APPROVAL PROCESS: CLINICAL AND REGULATORY MANAGEMENT, *edited by Richard A. Guarino*

Volume 31. TRANSDERMAL CONTROLLED SYSTEMIC MEDICATIONS, *edited by Yie W. Chien*

Volume 32. DRUG DELIVERY DEVICES: FUNDAMENTALS AND APPLICATIONS, *edited by Praveen Tyle*

Volume 33. PHARMACOKINETICS: REGULATORY • INDUSTRIAL • ACADEMIC PERSPECTIVES, *edited by Peter G. Welling and Francis L. S. Tse*

Volume 34. CLINICAL DRUG TRIALS AND TRIBULATIONS, *edited by Allen E. Cato*

Volume 35. TRANSDERMAL DRUG DELIVERY: DEVELOPMENTAL ISSUES AND RESEARCH INITIATIVES, *edited by Jonathan Hadgraft and Richard H. Guy*

Volume 36. AQUEOUS POLYMERIC COATINGS FOR PHARMACEUTICAL DOSAGE FORMS, *edited by James W. McGinity*

Volume 37. PHARMACEUTICAL PELLETIZATION TECHNOLOGY, *edited by Isaac Ghebre-Sellassie*

Volume 38. GOOD LABORATORY PRACTICE REGULATIONS, *edited by Allen F. Hirsch*

Volume 39. NASAL SYSTEMIC DRUG DELIVERY, *Yie W. Chien, Kenneth S. E. Su, and Shyi-Feu Chang*

Volume 40. MODERN PHARMACEUTICS, Second Edition, Revised and Expanded, *edited by Gilbert S. Banker and Christopher T. Rhodes*

Volume 41. SPECIALIZED DRUG DELIVERY SYSTEMS: MANUFACTURING AND PRODUCTION TECHNOLOGY, *edited by Praveen Tyle*

Volume 42. TOPICAL DRUG DELIVERY FORMULATIONS, *edited by David W. Osborne and Anton H. Amann*

Volume 43. DRUG STABILITY: PRINCIPLES AND PRACTICES, *Jens T. Carstensen*

Volume 44. PHARMACEUTICAL STATISTICS: PRACTICAL AND CLINICAL APPLICATIONS, Second Edition, Revised and Expanded, *Sanford Bolton*
Volume 45. BIODEGRADABLE POLYMERS AS DRUG DELIVERY SYSTEMS, *edited by Mark Chasin and Robert Langer*
Volume 46. PRECLINICAL DRUG DISPOSITION: A LABORATORY HANDBOOK, *Francis L. S. Tse and James J. Jaffe*
Volume 47. HPLC IN THE PHARMACEUTICAL INDUSTRY, *edited by Godwin W. Fong and Stanley K. Lam*

Additional Volumes in Preparation

HPLC in the Pharmaceutical Industry

edited by
Godwin W. Fong
SmithKline Beecham Pharmaceuticals
King of Prussia, Pennsylvania

Stanley K. Lam
Albert Einstein College of Medicine
of Yeshiva University
Bronx, New York

Marcel Dekker, Inc.   New York · Basel · Hong Kong

Coventry University

ISBN 0-8247-8499-5

P00642
12|8|97.

This book is printed on acid-free paper.

Copyright © 1991 by Marcel Dekker, Inc. All Rights Reserved.

Neither this book nor any part may be reproduced or transmitted in any form or by any means, electronic or mechanical, including photocopying, microfilming, and recording, or by any information storage and retrieval system, without permission in writing from the publisher.

Marcel Dekker, Inc.
270 Madison Avenue, New York, New York 10016

Current printing (last digit):
10 9 8 7 6 5 4 3 2

PRINTED IN THE UNITED STATES OF AMERICA

Preface

Although modern high performance liquid chromatography (HPLC) became popular in 1969, it was not widely accepted by pharmaceutical analysts until several years later. Four types of chromatography (namely, column, gas, paper, and thin layer) had been well established and officially employed in the isolations and assays of drugs as described in the United States Pharmacopeia (USP). However, the usefulness of HPLC techniques for pharmaceutical analysis was not appreciated by many practitioners in the pharmaceutical industry until the first HPLC systems capable of quantitative analysis became commercially available.

The acceptance of HPLC over gas chromatography in pharmaceutical analysis is mainly due to its improved selectivity and efficiency in separating nonvolatile organic drug molecules. Gas chromatographic techniques normally require a chemical derivatization step prior to the analysis of the drug, and chemical modifications of an organic drug molecule are undesirable in the stability studies of drug substances and drug products. HPLC techniques, on the other hand, offer enhanced detection sensitivity, improved accuracy, and reproducibility of drug analysis in the course of drug research, development, and quality control testing of marketed drug products. Many HPLC methods have been developed as the preferred methods for monitoring drug stability, identity of drug and other components, impurities, and degradation products of new chemical entities (NCE) or new molecular entities. The term "new molecular entities" has been used in the last several years when peptide and protein drugs discovered via biotechnology routes have been included. On the other hand, many wet chemistry and classical test methods for existing drug products have also been replaced by HPLC methods for more accurate measurements, better precision, and much faster analytical run time. This translates into lower cost per test in R&D and QC laboratories.

Automation, integrating HPLC with powerful data acquisition and reduction systems, and laboratory robotics have further improved the precision and accuracy of drug analysis by HPLC. Recognizing the reliability of the technique, both the pharmaceutical industry (as represented by such trade association as the Pharmaceutical Manufacturers Association [PMA]) and the regulatory agencies (such as the Food and Drug Administration [FDA]) have endorsed HPLC as the preferred methodology for testing drug samples. Indeed, PMA has drafted guidelines for the validation of HPLC methods intended for the assays of drugs and related compounds. FDA has been raising its standards of acceptable test results as scientific advances in the field of drug analysis move rapidly forward. Automated HPLC systems have been indispensable for any laboratories charged with the mandate of producing results that meet the "quality and quantity" standard set by company management.

This volume, composed of a total of 11 chapters grouped into four parts, was written by 18 experts in the field. Part One, Contemporary LC Techniques in Pharmaceutical Analysis, reviews the use of microbore and high speed LC and column switching techniques for a wide range of drugs. Part Two, Specialized Detection Techniques, covers electrochemical, radiochemical, and computerized diode array detection and HPLC/Fourier transform infrared (FTIR) for the analysis of drugs and their degradation products in formulations, and drugs and their metabolites in biological fluids. Part Three, Automation in Pharmaceutical Analysis, surveys the application of HPLC to the dissolution of solid dosage forms and robotic automation of HPLC. Part Four, HPLC of Peptides, Proteins, and Enantiomeric Drugs, includes the analysis of new drug substances and enantiomeric drugs using chiral HPLC techniques, and the characterization of peptide and protein drugs by HPLC.

This book could not have been completed without the dedication and commitment of our contributors. The encouragement and continued support of the publisher's staff and that of many concerned and knowledgeable colleagues is acknowledged. The input of the outside reviewers and of the series editor is very much appreciated. Last, but not the least, our heartfelt appreciation is expressed to each of our families for their patience, understanding, and support.

Godwin W. Fong
Stanley K. Lam

Contents

Preface   iii
Contributors   vii

Part One
Contemporary LC Techniques in Pharmaceutical Analysis

1. High Speed HPLC Using Short Columns Packed with 3 μm Particles   3
   Paul Kucera and Nicholas Licato

2. Microbore HPLC in Pharmaceutical Analysis   25
   Thomas V. Raglione and Richard A. Hartwick

3. Column Switching Techniques in Pharmaceutical Analysis   41
   Francis K. Chow

Part Two
Specialized Detection Techniques

4. Liquid Chromatography/Electrochemistry in Pharmaceutical Analysis   65
   Peter T. Kissinger and Donna M. Radzik

5. Radiochemical Quantitation: Considerations for HPLC   101
   Jeff Quint and John F. Newton

6. HPLC with Computerized Diode Array Detection in Pharmaceutical Research   123
   Ludwig Huber and H. P. Fiedler

7.	Design and Application of HPLC/FT-IR *Kathryn S. Kalasinsky and Victor F. Kalasinsky*	147

Part Three
Automation in Pharmaceutical Analysis

8.	Application of HPLC to Dissolution Testing of Solid Dosage Forms *William A. Hanson*	173
9.	Robotic Automation of HPLC Laboratories *Robin A. Felder*	185

Part Four
HPLC of Peptides, Proteins, and Enantiomeric Drugs

10.	Liquid Chromatographic Resolution of Enantiomers of Pharmaceutical Interest *Khanh H. Bui*	211
11.	HPLC of Proteins and Peptides in the Pharmaceutical Industry *Kalman Benedek and Joel K. Swadesh*	241

Index   303

Contributors

Kalman Benedek, Ph.D.[*] Pharmaceutics Department, Smith Kline & French Laboratories, King of Prussia, Pennsylvania

Khanh H. Bui, Ph.D Drug Disposition and Metabolism Department, ICI Pharmaceuticals Group, Wilmington, Delaware

Francis K. Chow, Ph.D. Quality Control, Par Pharmeceutical, Inc., Spring Valley, New York

Robin A. Felder, Ph.D. Department of Pathology, and Clinical Chemistry and Toxicology Laboratories, University of Virginia Health Sciences Center, Charlottesville, Virginia

H. P. Fiedler, Ph.D. Biological Institute, University of Tübingen, Tübingen, Federal Republic of Germany

William A. Hanson, Ph.D. Hanson Research Corporation, Chatsworth, California

Richard A. Hartwick, Ph.D.[†] Department of Chemistry, Rutgers University, Piscataway, New Jersey

Ludwig Huber, Ph.D. HPLC Detectors, Hewlett-Packard GmbH, Waldbronn, Federal Republic of Germany

Kathryn S. Kalasinsky, Ph.D.[‡] Mississippi State Chemical Laboratory, Mississippi State, Mississippi

---

Current affiliation:
[*]Terrapin Technologies, Inc., San Francisco, California
[†]Department of Chemistry, State University of New York at Binghamton, Binghamton, New York
[‡]Office of the Armed Forces Medical Examiner, Armed Forces Institute of Pathology, Washington, D.C.

Victor F. Kalasinsky, Ph.D.* Mississippi State University, Mississippi State, Mississippi

Peter T. Kissinger, Ph.D. Bioanalytical Systems, Incorporated, West Lafayette, Indiana

Paul Kucera, Ph.D. Analytical Development and Reference Standards, Lederle Laboratories, A Division of American Cyanamid Company, Pearl River, New York

Nicholas Licato, Ph.D. Analytical Development, Lederle Laboratories, A Division of American Cyanamid Company, Pearl River, New York

John F. Newton, Ph.D. Drug Metabolism Department, SmithKline Beecham Pharmaceuticals, Swedeland, Pennsylvania

Jeff Quint, Ph.D. Beckman Instruments Inc., Fullerton, California

Donna M. Radzik, Ph.D. Analytical Development, Lederle Laboratories Division, American Cyanamid Company, Pearl River, New York

Thomas V. Raglione, Ph.D.† Department of Chemistry, Rutgers University, Piscataway, New Jersey

Joel Swadesh, Ph.D.‡ Analytical Chemistry Department, Smith Kline & French Laboratories, King of Prussia, Pennsylvania

---

Current affiliation:
*Department of Environmental and Toxicologic Pathology, Armed Forces Institute of Pathology, Washington, D.C.
†Bristol-Myers Squibb Corporation, Somerset, New Jersey
‡Department of Veterinary and Animal Sciences, University of Massachusetts at Amherst, Amherst, Massachusetts

HPLC in the
Pharmaceutical Industry

Part One
Contemporary LC Techniques in Pharmaceutical Analysis

1

High Speed HPLC Using Short Columns Packed With 3 μm Particles

Paul Kucera and Nicholas Licato

Lederle Laboratories, A Division of American Cyanamid, Pearl River, New York

HISTORY OF HIGH SPEED CHROMATOGRAPHY

The history of high speed chromatography goes back to 1959, when Purnell and coworkers [1] first attempted to construct high speed gas chromatographic columns. In those days, relatively little was known of the approach to higher analytical speeds. In his article "An Approach to Higher Speeds in Gas-Liquid Chromatography," Purnell correctly described for the first time the retention time optimization and showed that the speed of the analysis is inherently dependent on the shape of the plate height function, particularly the van Deemter mass transfer coefficients of the solute transfer between the mobile and the stationary phase. Gas chromatography was considered a much faster technique than liquid chromatography, mainly because it was realized that diffusion rates of substances in gas media are four to five orders of magnitude greater than in any liquid systems. Thus any attempts to speed up liquid chromatographic separations were completely neglected.

In 1974, however, when small particles and column packing systems and procedures utilizing high pressure were developed for liquid chromatography, various papers appeared in the literature describing the time optimization [2-10], and since that time public awareness of high speed liquid chromatographic systems has greatly increased. Perhaps the main reason for this new renaissance in liquid chromatography is the fact that the speed of the analysis directly affects the economy and the operating cost of the analysis. It was realized that the optimization of a chromatographic system consists of optimizing the four basic attributes: resolution, speed, load, and scope [11]. The scope of the system was generally considered as the system capacity for separation mixtures of wide polarity range, and it was usually varied by using the technique of gradient elution in liquid chromatography or tem-

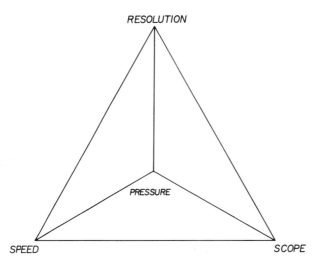

Figure 1 Tetrahedron of basic chromatographic attributes.

perature programming in gas chromatography. The load was directly related to column cross-sectional area or column diameter and made all the difference between the preparative systems and analytical separations. For purely analytical chromatography, which will be discussed in this chapter, the load can be replaced with column operating pressure. A typical chromatography tetrahedron with main operating attributes of resolution, speed, scope, and pressure is shown in Figure 1. Attempts to reduce the pressure drop on the column further led to the development of high speed open tubular capillary systems and was allowed to optimize based on the column or particle diameter. It is now well understood that the increase in the resolution of a chromatographic system can be achieved by obtaining high efficiency in terms of plate numbers, long column lengths, and high selectivity, while improvements in speed can be attained by using small particles and short columns operated at high mobile phase velocities. As shown in Figure 2, the chromatographic resolution is of primary importance for any meaningful analytical work, while speed is of secondary importance and tied up to the economy of the operation.

Desty contributed significantly to high speed systems by suggesting the use of theoretical plates or effective theoretical plates per second as a criterion of speed in rapid analyses. Shortly after it was demonstrated that open tubular capillary columns led in speed, Desty showed that 2000 plates/sec can be achieved in gas chromatography using specialized equipment. In 1979, a 25 sec separation of 8 component mixture exhibiting about 600 plates/sec was published by Scott, Kucera, and Monroe [12].

The chromatogram shown in Figure 3 was obtained on a microbore 25 cm × 1.0 mm id column packed with 20 μm silica gel and operated at a 4.5 mL/min flow rate. This chromatogram signaled the advent of a new era in chromatography, because

High Speed HPLC Using Short Columns

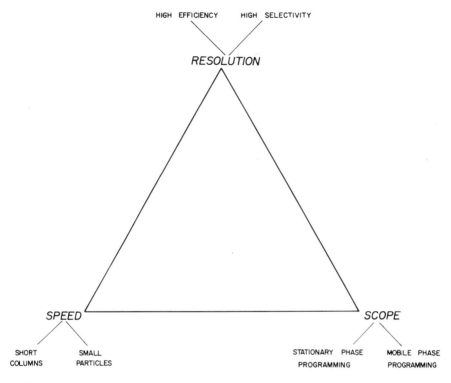

Figure 2 Operative considerations of scope, speed, and resolution.

besides the extremely high speed of separation, one never achieved before in liquid chromatography, the economy of analysis was significantly increased by using a microbore column system. As will be seen later, high speed separations theoretically require short columns packed with small particles. Unfortunately, the column pressure drop increases with the reciprocal of the square of the particle diameter, and thus the process of reducing the particle diameter in a column cannot be carried out ad infinitum. From the work of Guiochon [13], it appears that approximately 2 µm is the optimum particle diameter for a packed column. Beyond this limit, any further improvements in the column performance would be impaired by excessively high pressure exerted on the column. It is interesting to note that currently there are not many 3 µm particle columns on the market, and virtually no 2 µm packed columns are available. The majority of commercially available columns are packed with 5–8 µm spherical particles. Perhaps the reasons for the lack of high efficiency columns packed with small particles are the increasing difficulty of packing these small particles into long columns and the quality of the particles. The following discussions in this chapter will focus mainly on high speed liquid chromatographic systems.

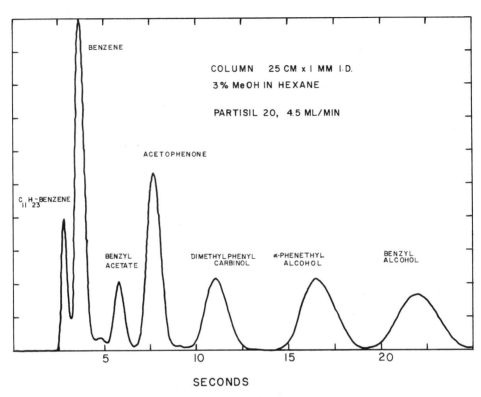

Figure 3 Rapid and complete separation of a 7 component mixture on a 25 cm × 1.0 mm id column packed with Partisil-20. 3% methanol in pentane:hexane (50:50) mobile phase at 4.5 mL/min flow rate.

THEORETICAL ASPECTS OF HIGH SPEED SYSTEMS

The effective use of the liquid chromatographic column requires that the separation of two substances A and B be achieved within a desired resolution R, and that N number of theoretical plates are needed to achieve this goal. Thus at the start of the time optimization procedure, it is useful to define the value of R needed by

$$R = \frac{\sqrt{N}\,(\alpha - 1)k'_B}{4\alpha(1 + k'_B)} \tag{1}$$

It should be noted that the k' is the familiar solute capacity factor defined by

High Speed HPLC Using Short Columns

$$k' = \frac{t_r - t_0}{t_0} \qquad (2)$$

where t_r is the solute retention time and t_0 is the retention time of the unretained solute. The selectivity α is typically defined as the ratio of two capacity factors; $\alpha = k'_B/k'_A$. Furthermore, the solute retention time depends on the solute capacity factor k' and the column length by

$$t_r = t_0(1 + k') = \frac{L}{u}(1 + k') = \frac{HN}{u}(1 + k') \qquad (3)$$

where L is the column length, u is the linear velocity of the system, and H is the so-called HETP or the height equivalent to theoretical plate. Substituting for N from Equation (1) to Equation (3) will yield the retention time function:

$$t_r = 16R^2 \left(\frac{\alpha}{\alpha - 1}\right)^2 \frac{(1 + k'_B)^3}{k'^2_B} \frac{H}{u} \qquad (4)$$

where H/u is the slope of the HETP curve, which generally depends on the particle diameter of the packing material. Many workers have demonstrated that the smaller the particle diameter of the packing material the smaller the H/u ratio. Typical examples of HETP curves obtained with a microbore column 1 m in length and 1.0 mm id for 5, 10, and 20 μm particles are shown in Figure 4. As can be observed from Figure 4, for retained solute ($k' = 1.8$), the smaller particle diameter packing results in a decrease in the slope of H/u, flattening the plate height function curve. The result is that less chromatographic band deterioration will be observed at higher speeds for smaller diameter particles.

From Equation (4), the absolute value of H depends on the particle diameter, and this feature makes comparison of columns packed with different particles difficult. In order to overcome this problem, Giddings [14] introduced the concept of reduced plate height* h where

$$h = \frac{H}{d_p}$$

and reduced velocity ν defined as

---

*Reduced plate height is usually described by the so called Knox equation; $h = B/\nu + A\nu^{1/3} + C\nu$ where A, B, and C are constants.

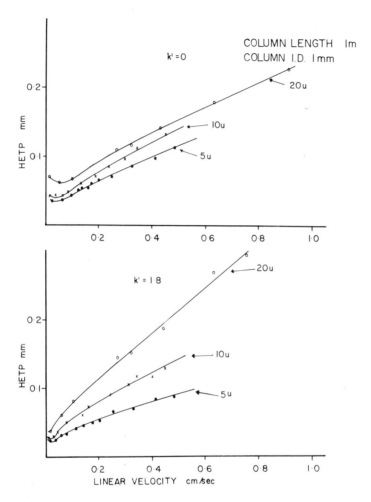

Figure 4 HETP curves for microbore columns packed with silica gel having different particle diameters.

$$v = \frac{ud_p}{D_m}$$

where d_p is the particle diameter of the packing material and D_m is the solute diffusion coefficient in the mobile phase. From Equation (4), t_r can be expressed as a function of α, k′, and H/u ratio. After introduction of the reduced parameters and assuming that the values of k′ and α are fixed, one gets a direct dependence of solute retention time on the particle diameter of the packing material and solute diffusivity.

$$t_r = f(k', \alpha, H/u)$$

$$t_r = \text{constant} \, \frac{hd^2}{\nu D_m}$$

Therefore it is easily understood why high speed chromatography systems should utilize short columns packed with small particles operated at high mobile phase velocities.

A careful examination of Equation (4) shows that the speed of analysis depends on three basic factors:

$$\frac{(1 + k'_B)^3}{k_B^{'2}} \quad \text{the k' parameter}$$

$$\frac{\alpha^2}{(\alpha - 1)^2} \quad \text{the selectivity parameter}$$

$$\frac{H}{u} = \frac{hd_p^2}{\nu D_m} \quad \text{the plate height parameter}$$

Generally, the speed of the analysis is related to the retention time of the last eluting component n on the chromatogram, $(t_r)_n$. Therefore the speed is

$$S = \left[(t_r)_n \left(1 + \frac{2}{\sqrt{N}} \right) \right]^{-1}$$

and the faster the speed the shorter the analysis time. Thus for a highly efficient high speed liquid chromatographic system, optimum values of k′, α, and H/u are needed. The optimum k′ value can be calculated as follows, assuming that H/u and selectivity are constant:

Table 1 Equations of Resolution, Retention Time, Theoretical Plates, and Plate Height in Liquid Chromatography

Resolution equation	$R = \dfrac{\sqrt{N}}{4} \dfrac{k'}{(1+k')} \dfrac{(\alpha-1)}{\alpha}$	(A.1)
Retention time equation	$t = \dfrac{L}{u}(1+k') = \dfrac{NH}{u}\left(1 + K\dfrac{V_s}{V_m}\right)$	(A.2)
Theoretical plate equation	$N = 16\left[\dfrac{\alpha}{\alpha-1} \dfrac{(1+k')}{k'}\right]^2$	(A.3)
Retention time equation (reduced parameters)	$t = 16\left[\dfrac{\alpha}{\alpha-1}\right]^2 \dfrac{(1+k')^3}{K'^2} \dfrac{hd^2}{\nu D_m}$	(A.4)
Knox equation	$h = \dfrac{B}{\nu} + A\nu^{1/3} + C\nu$	(A.5)

$$\frac{dt_r}{dk'} = 0 = C' \frac{k'^3 - 3k' - 2}{k'^3}$$

This equation leads to the polynomial function

$$k'^3 - 3k' - 2 = 0$$

which has a solution at k' = 2. To express this result differently, a high speed chromatographic system should not use excessively high capacity factors, and the k' value for the last eluting peak should not exceed 2. However, it should be realized that a prerequisite for any chromatographic system is the separation of two substances of interest, and thus for this purpose the column selectivity (or the so-called selectivity parameter) is of primary importance. The summary of equations used in the previous time optimization procedures is shown in Table 1.

It would seem that the smaller the particle diameter the faster the chromatographic system, and this is certainly true up to a certain limit. Unfortunately, the column pressure drop ΔP is proportional to the reciprocal of the square of the particle diameter, and thus the column pressure increases drastically with reduction in particle size:

$$\Delta P = \frac{Lu\eta}{k_0 d_p^2}$$

In the above equation k_0 is the specific permeability, which typically has a value of about 0.001 for a random dense packing used in HPLC columns, and η is the solvent viscosity. It can be seen that successful optimization of a high speed system also requires the use of solvents of low viscosity and again short column length to compensate for the effect of high pressure generated as a result of small particle size. The column pressure is a limiting factor, because the practical pressure rating of the injection and pumping system is about 6000 psi (about 414 atm). The specific column permeability is another limitation, because for a well packed column this value is virtually constant, unless open tubular capillary systems are used; in this instance k_0 = 1/32, and a faster speed of analysis can be achieved. It can be seen that the optimization of a high speed chromatographic system is a complex issue that depends on many parameters. Beside the theoretical aspects of the optimization process, one cannot overlook also the experimental aspects of high speed operation, which will be discussed next.

DESIGN OF INSTRUMENTATION FOR HIGH SPEED SEPARATIONS

All chromatographic systems are essentially entropy driven processes. Assuming that a certain amount of energy is introduced into the system, the separation of sample components from the mixture of given enthalpy and entropy results in the reduction of entropy and, typically, increase in enthalphy of the system, which dissipated as heat of adsorption. Unfortunately, this nonideal, irreversible process is also accompanied by an increase in entropy due to dilution of each sample component in the mobile phase and band broadening of the solute bands as they travel through the injector, column, capillary tubing, and detector and become distorted due to associated detector electronics. A chromatographic system for high speed separations must therefore be uniquely designed to minimize the band spreading processes in the chromatograph.

Let us assume that the solute elution profile can be described by a symmetrical Gaussian function and that any individual band dispersion process in the system is independent of any other process. The total variance σ_T^2 of the band observed at the detector can be then expressed as the sum of all variances:

$$\sigma_T^2 = \sigma_S^2 + \sigma_C^2 + \sigma_D^2 + \sigma_t^2 + \sigma_e^2$$

where

σ_S^2 = the volume variance due to sampling,
σ_C^2 = the variance of the solute band leaving the column,
σ_D^2 = the variance resulting from the solute dispersion in the detector cell,
σ_t^2 = the variance due to connecting tubing, and
σ_e^2 = the variance due to detector and data processing electronics.

Let us consider each of these dispersion processes and how they can be minimized for high speed chromatographic separations.

Sample Variance

The theoretical value of the sample variance has been calculated from a second statistical moment of rectangular distribution as well as experimentally determined by various workers [15,16]. Guiochon proposes that the band variance of the sample should be proportional to the square of the sample volume V_S.

$$\sigma_S^2 = \frac{V_S^2}{K^2}$$

where the constant K depends on the injection technique; under ordinary experimental conditions $K^2 = 4$. The theoretical value for K^2 is 12, which is difficult to achieve in experimental work.

The constant K typically depends on sample volume, speed of the sample application on the column, internal dead volume of the sampling system device, and flow pattern in the injector. The maximum sample volume V_{SM} has been calculated [17] as a function of column length and column and particle diameter.

$$V_{SM} = 0.34 \sqrt{L d_p d_c}$$

High speed injection systems operated at high mobile phase velocities and high pressures should incorporate very fast sample switching intervals (of a few milliseconds) in order to reduce band spreading in the sampling system and prevent sudden pressure buildup on the column or injector.

Column Variance

The variance due to the column depends on column dead volume V_0, column efficiency N, solute capacity factor k', and indirectly on the flow rate Q of the mobile phase through the column:

$$\sigma_C^2 = \frac{V_0(1 + k')}{\sqrt{N}} = \frac{Q t_0(1 + k')}{\sqrt{N}}$$

High Speed HPLC Using Short Columns

As described previously, high speed chromatographic systems require short columns packed with small particles and operated with high mobile phase velocities. It is the unfortunate property of all chromatographic systems that the deterioration of system performance increases with higher flow or velocity of the mobile phase. Thus in order to convert an existing analytical liquid chromatograph for rapid separation one must trade off the column efficiency and performance for higher speed of analysis. In practice, the highest column efficiency and performance must be selected for rapid separations; otherwise the system would not be practical.

Detector Variance

A significant band dispersion can result from a poorly designed detector cell. A similar consideration carried out for the sampling system can be also applied to a typical flow-through detection system. For a concentration sensitive detector, such as a UV detector, the detector volume variance will depend on the detector cell volume and the experimental constant M:

$$\sigma_D^2 = \frac{V_D^2}{M^2}$$

The volume variance can be converted to a time variance as shown below; Q is the mobile phase flow rate:

$$\sigma_{D,t}^2 = \frac{V_D^2}{Q^2 M^2}$$

Various authors quote practical considerations and rules that allow the determination of relationships between the detector cell volume and related parameters such as band volume, maximum volume, and detector time or volume variance.

For example, Kirland [18] has shown that if the detector cell volume is smaller than 1/10 of the peak volume, the extra column band broadening resulting from the detector cell is insignificant. A practical rule observed by many chromatographers states that the detector cell volume should be equal to about one third of the maximum sample volume as calculated previously.

Connecting Tube Variance

It is presently well understood that capillary tubing connections between the detector, column, and injector can cause significant band spreading in a chromatographic system. This contributes to poor column performance, particularly for a high speed LC system. Assuming an open tubular capillary tubing, the connecting tube variance contribution has been theoretically determined [19] and found to be dependent on the flow rate of the mobile phase, column length and diameter, and solute diffusion coefficient D_m:

$$\sigma_t^2 = \frac{D^4 \pi L Q}{384 D_m}$$

Therefore connecting tubes of very short length and diameter should be employed for high speed systems. However, great care in designing connecting tubes for high speed systems should be taken, because the pressure drop increases with the decrease in tubing diameter. This feature also calls for somewhat higher pressure ratings of the detector cell and connecting fittings.

Detector Electronics and Data Acquisition Variance

Finally, a significant amount of solute band spreading can be obtained from various detector components such as detector photosensor, amplifier, time constant capacitances, and appropriate data acquisition electronics. It should be understood that although the actual solute band passing through the detector cell will not be dispersed or further diluted, the final chromatographic record will show a distortion of the peak due to the above process. It has been discussed previously that due to a requirement for high volumetric flow needed in rapid separation systems, most practical and economical systems should be based on small diameter microbore columns. However, since the absolute band volume eluted at high speed from a column of small diameter is very small (typically a few microliters), and the band is sweeping the detector cell at high velocity, the response of the detector photosensor (either photomultiplier or cadmium sulfide photoresistor) must be also very rapid. In addition, any band distortion resulting from the time constant of the amplifier, the time constant of the recorder or plotter, and the time constant of the computer data acquisition equipment must also be minimized.

The volumetric column variance σ_C^2 of the solute band leaving the column can be converted to standard solute time variance of the column $(\sigma_C^2)_t$ as follows:

$$(\sigma_C^2)_t = \left[\frac{V_0(1+k')}{\sqrt{N}Q} \right]^2 = \frac{\sigma_C^2}{Q^2}$$

The time variance T^2 as seen on the recorder or plotter then consists of the column time variance plus the time variance due to the electronics of the detection/recording system $(\sigma_e^2)_t$.

$$T^2 = (\sigma_C^2)_t + (\sigma_e^2)_t$$

Assuming that a 5% increase in the solute band variance can be tolerated as a result of the above electronic distortion processes,

$$T = 1.05(\sigma_C)_t$$

and

$$T^2 = (1.05)^2(\sigma_C^2)_t = (\sigma_C^2)_t + (\sigma_e^2)_t = 1.103(\sigma_C^2)_t$$

Thus

$$(\sigma_e^2)_t = 0.103(\sigma_C^2)_t$$

and the total standard deviation due to electronic distortion processes can be calculated as

$$(\sigma_e)_t = 0.32(\sigma_C)_t$$

Based on these considerations, the peak distortion due to the time constants of the detector, recorder, etc. that can be tolerated is about 32% of the time standard column deviation in seconds. For example, for a 30 sec high speed separation of 5 components, a time constant of about 0.2 sec is needed. Unfortunately, the total detector time constant of many detectors is currently greater than 0.5 sec and this feature limits the speed of analysis and resolution that can be attained with commercial equipment.

It can be seen from the previous considerations that the high speed liquid chromatographic system has to be very carefully designed in order to prevent the solute band from spreading in the system. It is hoped that in the future, specialized detectors, sampling devices, and columns will be available from instrument manufactures that will permit achievement of the high column efficiencies and performances while operating the chromatographic systems in a high speed mode.

APPLICATIONS OF HIGH SPEED ANALYSIS

Microbore columns packed with small particles have achieved high speed separations unattainable previously. Because of the economy of scale, microbore methods are extensively found in quality control applications. And because of the speed, microbore methods are widely used in clinical and basic research. The separation of Diazepam from its metabolites, using a column 1 × 250 mm packed with Partisil-10, 10 µm particles, is shown in Figure 5. Baseline resolution of the drugs was accomplished in less than 45 sec with a mobile phase flow rate of 1 mL/min. This example demonstrates the separation of a microsample of a complex mixture on a moderately efficient microbore column with the mobile phase consumption scaled down to a minimum.

Derivatization with O-phthadialdehyde is one of many ways to enhance the sensitivity of detection of primary amines and amino acids. The reaction is mild, rapid,

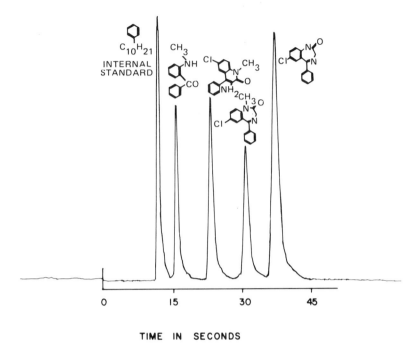

Figure 5 The separation of diazepam and its metabolites on a microbore column in 43 sec. 25 cm × 1.0 mm id column packed with Partisil-10. 8% methanol and 10% ethyl acetate in n-hexane mobile phase at 1 mL/min flow rate.

and complete, giving reproducible yields. Since the resulting adduct is highly fluorescent, a small sample is sufficient. The problem of limited sample quantities, as with amino acid compositions in protein hydrolysates, can be easily addressed by precolumn derivatizing with OPA and separating the derivatives on microbore columns. As shown in Figure 6, a representative mixture of amino acids was resolved in 20 min on a 1 × 500 mm Zorbax column with 8 μm ODS packings, at a flow rate of 38 μL/min. The detection limit of the amino acids, with a Kratos SF 979 fluorescence detector set a 418 nm with excitation at 330 nm, was 100 femtomoles at two time signal to noise.

While precolumn derivatization decreases bandspreading of the chromatographic system, nevertheless with little sacrifice in efficiency the microbore system can be modified to accommodate postcolumn derivatization of primary amines with OPA. As shown in Figure 7, the separation of the aliphatic amines on a 1 × 500 mm Alltech column with 7 μm ODS materials in 20 min with postcolumn derivatization was achieved. Each peak in the chromatogram represents 40 pmoles of amine in a 0.2 μL in injection.

High Speed HPLC Using Short Columns

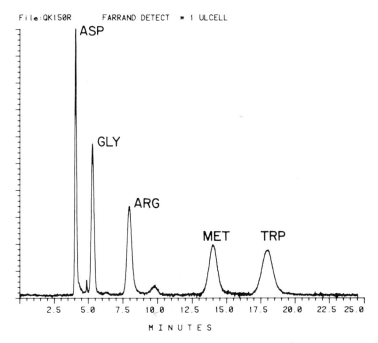

Figure 6 The separation of OPA amino acids using precolumn derivatization. Zorbax ODS 8 μ, 50 cm × 1 mm id. 65% 0.01 M NaAc buffer, 38 μl/min flow rate. Excitation wavelength 330 nm; emission wavelength 418 nm.

As discussed in the theoretical consideration, columns packed with smaller particles give higher efficiency. Using a 75 × 46 mm Ultrasphere column with 3 μm ODS materials at a flow rate of 1.5 mL/min, a complex mixture of amino acids was separated as the OPA derivatives in 50 min as shown in Figure 8. Faster separations were realized with 100 × 4.6 mm columns packed with 3 μm ODS Ultrasphere and Microsorb packings as shown in Figures 9a and 9b respectively [19]. With the 3 μm material, the analysis time was reduced and the resolution was improved over the corresponding 5 μm packing material.

The chromatographic fingerprinting of amino acids shown in Figure 9 was applied to protein hydrolysates of carboxymethylated calcitonin. Figure 10a shows the acid hydrolysate and Figure 10b the aminopeptidase digest of the protein separated under the identical conditions ad Figure 9a and 9b. Good agreement between these profiles and the reported amino acid sequence of human calcitonin was noted. This exemplifies the resolution, speed, and sensitivity afforded by 3 μm high speed columns when coupled to fluorescence derivatization techniques.

The application of microbore columns is extended to nucleotide analysis. Using a 1 × 250 mm Partisil column with a 10 μm ODS material, ribonucleotide dimers were

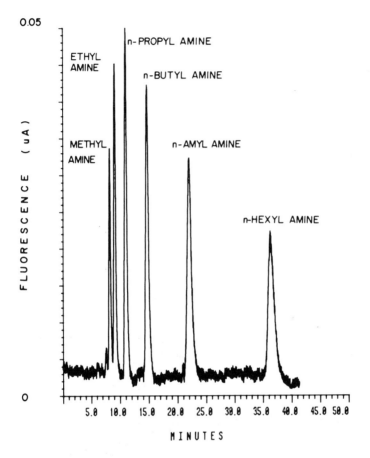

Figure 7 The separation of aliphatic amines on a microbore column using a postcolumn fluorescence derivatization system. Alltech ODS 7 μ, 50 cm × 1 mm id. 38% v/v acetonitrile/ 0.01 M KH_2PO_4, flow rate (column and reagent, each) 35 μL/min. Reactor volume of 44 μl, 175 cm × 180 μm id stainless steel tubing. Detector cell volume of 0.25 μl.

separated with a mobile phase containing 20% methanol in 0.01M KH_2PO_4 at a flow rate of 50 µL/min as shown in Figure 11. Since oligonucleotide fragments are ordinarily prepared on a small scale, microbore chromatography can yield purified materials suitable for successive synthesis and continued analysis. Deoxyribonucleotides were resolved under similar chromatographic conditions using approximated 2 mg of each solute as appeared in Figure 12 [5].

Microbore methods, being fast and sensitive, are popular in biomedical applications. Developments in high speed chromatography promise a future with improved column technology and highly sensitive ultratrace detection methods.

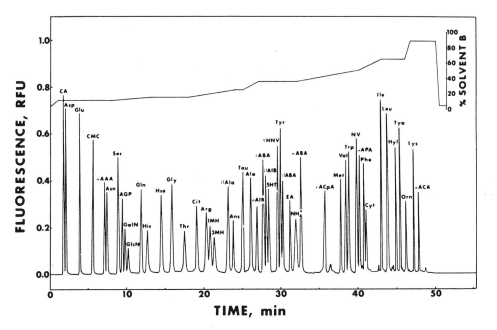

Figure 8 Elution profile of an OPA-derivatized standard mixture of amino acids and biological amines chromatographed on a 2 µ particle size Ultrasphere ODS column (75 × 4.6 mm id). Each peak represents 80 picomoles except for ammonia at 330 picomoles.

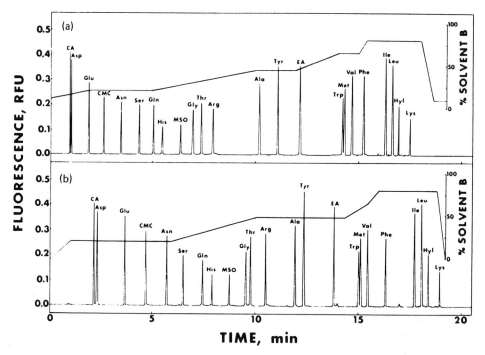

Figure 9 Elution profile of OPA-derivatized amino acids chromatographed on 3 µ particle size reversed phase columns. (a) Ultrasphere ODS column (100 × 4.6 mm id); flow rate 1.7 mL/min. (b) Microsorb C-18 column (100 × 4.6 mm id); flow rate 1.6 mL/min. Each peak represents 20 picomoles.

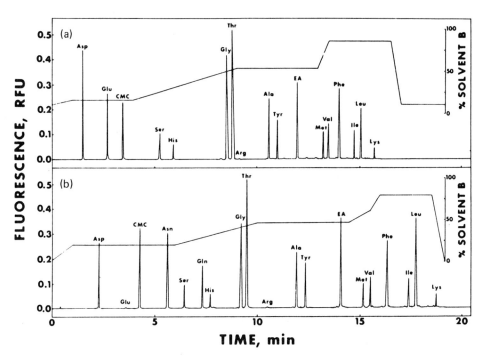

Figure 10 Elution profiles of hydrolysates of carboxymethylated human calcitonin. Conditions for (a) and (b) as given in Figure 9.

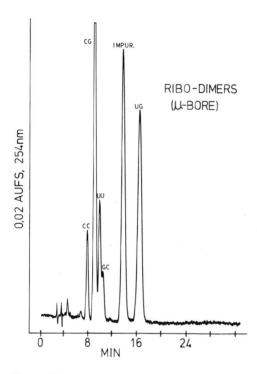

Figure 11 Separation of dimers of the ribonucleotide monophosphates using a packed 1 mm id microbore column. Whatman glass lined, 25 cm × 1 mm id. 0.01 KH_2PO_4, pH 5.6 with 20% methanol, 50 µL/min flow rate. Injection volume of 0.5 µL of 1 mg/mL in each of the dimers.

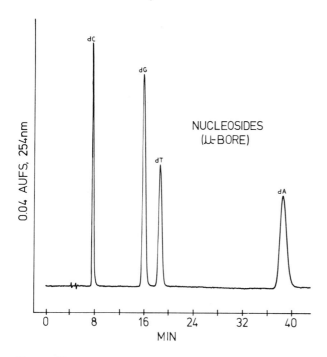

Figure 12 Deoxyribonucleosides and their mononucleotides separated on a set of 2 Whatman microbore columns coupled for greater efficiency. Approximately 2 picograms of each solute injected. Whatman Partisil-10 ODS, 0.01 KH_2PO_4, pH 5.6 with 20% methanol, 50 µL/min flow rate.

REFERENCES

1. D. H. Desty in J. C. Giddings and R. A. Keller (ed.), *Advances in Chromatography*, Marcel Dekker, New York, Vol. 1, p. 100 (1965).
2. G. Gasper, C. Vidal-Madjar, and G. Guiochon, *Chromatographia, 15*: 125 (1982).
3. J. L. DiCesare, M. W. Dong, and L. Ettre, *Chromatographia, 14*: 257 (1981).
4. J. L. DiCesare, M. W. Dong, and J. G. Atwood, *J. Chromatogr., 217*: 369 (1981).
5. R. A. Hartwick and D. D. Dezaro in *Microcolumn High-Performance Liquid Chromatography* (P. Kucera, ed.), Elsevier, New York, p. 75 (1984).
6. C. E. Reese and R. P. W. Scott, *J. Chromatogr. Sci., 18*: 479 (1980).
7. G. Guiochon, *Anal. Chem., 53*: 1381 (1981).
8. R. P. W. Scott and P. Kucera, *J. Chromatogr., 169*: 51 (1979).
9. Z. Yukuei, *J. Chromatogr., 197*: 97 (1980).
10. J. J. Kever, *J. Chromatogr., 207*: 145 (1981).
11. R. P. W. Scott and P. Kucera, *J. Chromatogr. Sci., 12*: 473 (1974).
12. Scott, Kucera, and Monroe, *J. Chromatogr., 186*: 475 (1979).
13. G. Guiochon, *Anal. Chem., 52*: 2002 (1980).
14. J. C. Giddings (ed.), *Dynamics of Chromatography*, Part I, Marcel Dekker, New York, 1965.
15. A. Klinkenberg, *Gas Chromatography*, Butterworth, London, p. 182 (1960).
16. J. F. K. Huber and J. A. R. Hulsman, *Anal. Chim. Acta, 38*: 305 (1967).
17. G. Guiochon in *Microcolumn High-Performance Liquid Chromatography* (P. Kucera, ed.), Elsevier, New York, p. 6 (1984).
18. J. J. Kirkland, *J. Chromatogr. Sci., 15*: 303 (1977).
19. B. N. Jones and J. P. Gilligan, *J. Chromatogr., 266*: 471 (1983).

2
Microbore HPLC in Pharmaceutical Analysis

Thomas V. Raglione* and Richard A. Hartwick†

Rutgers University, Piscataway, New Jersey

INTRODUCTION

Since the first paper by Scott and Kucera [1] popularizing narrow bore columns (that is, columns with diameters appreciably less than 2 mm id) appeared, interest in micro-LC has continued to increase. Small diameter columns had been used in the earliest days of HPLC [2–5] with pellicular packings, which because of their large particle size and low capacity had to be used with relatively long columns. The high efficiencies achieved by Scott and Kucera, however, nearly 750,000 plates per column [6], were the highest ever achieved at that time and are still quite remarkable today. At the other performance extreme, Scott and Kucera [7] and others [8] were able to separate and quantify various drug formulations within several seconds total analysis time by using short, optimized 1 mm id columns generating only several hundred theoretical plates.

These and other early results with micro-LC instrumentation initiated a wave of research aimed at understanding and developing small diameter columns. Though these first columns were indeed microbore in contrast to their large bore (4.6 mm) cousins, the trend to reduce column diameters even further soon outpaced column terminology. A recent paper has attempted to reorganize column terminology [9] to encompass narrow bore, packed capillary, and open tubular microbore columns. Today chromatographers have commercially available an entire spectrum of column diameters ranging from packed capillaries of several hundred micrometers up to preparative column diameters in the centimeter range. Though quantized in prac-

---

Current affiliations:
*Bristol-Myers Squibb Corporation, Somerset, New Jersey.
†State University of New York at Binghamton, Binghamton, New York

tice, the column diameter is yet another variable to be optimized in a separation by the working scientist.

ADVANTAGES OF SMALL BORE COLUMNS

Early reviews of microbore chromatography [10-13] listed four main advantages microbore columns have over conventional columns: (a) increased mass sensitivity (which will be defined and discussed later), (b) decreased solvent consumption, (c) high total column efficiencies, and (d) rapid thermal equilibration. In addition to these advantages, it is possible to add (e) the ability to directly connect the LC system to new detectors and (f) column scaling in multidimensional chromatography separations. Of all of these advantages, the direct interface of LC with the mass spectrometer is by far the most significant. LC/MS, having the potential to analyze a far larger number of compounds than GC/MS, will eventually become one of the analytical chemist's most powerful tools.

Micro-LC had been predicted to develop a substantial user base in pharmaceutical research laboratories [14,15]. However, this has not materialized, primarily due to the stringent dead volume requirements of these small bore columns, which often precludes their use with many existing autosamplers. Figure 1 illustrates the deleterious effect that a few microliters of dead volume can have on a microbore separation [16]. An additional reason for the apparent lack of interest in microbore HPLC by the pharmaceutical industry is the need for modification or replacement of existing instrumentation. Since many companies develop methods to be run worldwide at various laboratories, they need to insure that all of these laboratories have the instrumental capability of running the analyses.

Mass Sensitivity

One area of pharmaceutical analysis where micro-LC is particularly suited is the determination of impurities and metabolic pathways of drugs. Since the sample volumes of body fluids and tissues are often quite limited, microbore LC is well suited for the analysis of these samples due to its higher mass sensitivity as compared to conventional (4.6 mm id) columns.

The increased mass sensitivity of microbore columns has been an area of confusion in the literature. If one injects a sample that just gives a response (2 times the noise) on the microbore system and then injects the same volume of that sample on the conventional system, no peaks will be detected. In this situation, the microbore system is more mass sensitive, since the same mass was injected on both systems but detected only on the microbore system. But if the injection size is properly scaled to the column diameter, i.e., both columns have the same sample loading, µg of sample per mg of packing material, then the signal should be the same. Thus only when sample volume is limited is there any advantage in reducing the column diameter. It is questionable how many real analytical situations present themselves in which samples are so limited, although such cases do of course exist.

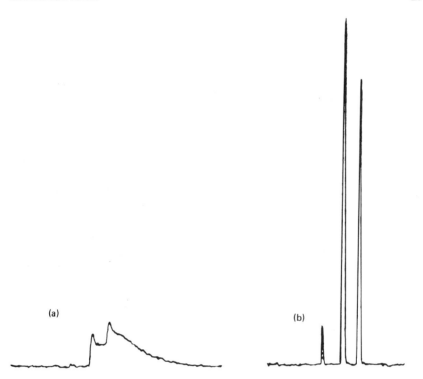

Figure 1 An illustration of the deleterious effects that a few microliters of dead volume can have on a separation. (a) dead volume ca. 2 µL; (b) minimum dead volume ca. 30 nL. Sample: benzene-toluene. Column: 25 cm × 0.22 mm id. 5 µm Nucleosil C. Mobile phase: MeOH-water (80:20). (Reproduced with permission from Ref. 13.)

The minimum detection level is defined as that amount of solute that is required to give a signal twice the noise level and is expressed in units of mass. This can be derived in the following form [17],

$$m = \frac{2Ld_C^2 X_D(1 + k')}{N} \quad (1)$$

where m is the porosity of a column of length L and diameter d_C having N plates, k' is the capacity factor of the solute of interest, and X_D is the concentration sensitivity of the detector. Equation (2) calculates the increase in mass sensitivity due to reduction in column diameter [9], by assuming equal column lengths, efficiencies, and porosities:

$$\text{reduction factor} = \frac{d_1^2}{d_2^2} \tag{2}$$

where d_1 and d_2 are the column diameters being examined. This decrease in the minimum detection limit is a theoretical limit and is rarely achieved in practice due to the inability of separating chromatographic from spectroscopic effects.

In pharmaceutical analysis the mass sensitivity advantage is rarely taken advantage of since the majority of analyses are not sample limited. Pharmaceutical chromatographers often have several grams of the sample at their disposal. However, the increase in mass sensitivity is quite relevant when dealing with trace work such as drug degradation or drug metabolism in tissues or cells. Takeuchi et al. [18] have successfully resolved 18 amino acids in only 1.3 nL of soya sauce (Figure 2), demonstrating the potential for chemical analysis of individual cells.

In drug degradation studies the increased mass sensitivity will allow the method development chemists to identify the degradation products more easily, though routine analyses would probably be done on conventional columns with scaled up sample sizes. In drug metabolism, the analyst is concerned with tracing the drug through

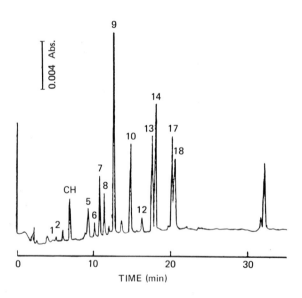

Figure 2 Separation of 18 amino acids in 1.3 nL soya sauce. Conditions: Column ODS-Hypersil (3 μm), 100 × 0.34 mm id. Mobile Phase acetonitrile-0.13 M ammonium acetate. Flow rate 4.2 μl/min; UV 222 nm. Key to compounds: (1) Asp, (2) Glu, (OH) dansylic acid, (5) Ser, (6) Thr, (7) Gly, (8) Ala, (9) Pro, (10) Val, (12) Met, (13) Ile, (14) Leu, (17) Phe, (18) NH. (Reproduced with permission from Ref. 15.)

Table 1 Solvent Consumption for Columns of Different Diameters[a]

D_c (mm)	Flow rate
4.6	1 mL/min
2.0	190 µL/min
1.0	47 µL/min
0.25	3.6 µL/min
0.075	0.27 µL/min

[a]Calculations are for a constant linear velocity of 1.4 cm/sec.

its metabolic pathway in some living organism; often this will require analysis of samples of very limited size.

Efficiency

Theoretically, there is no valid reason why 1.0 mm columns should be more efficient than 4.6 mm columns. In practice, however, it is much easier efficiently to pack long lengths of small bore columns than to pack conventional columns. Small bore columns, due to their geometry, are easier to couple with minimal loss in efficiency than conventional columns, generating columns of several hundred thousand total theoretical plates. Halasz [19], however, has recently demonstrated the successful coupling of a series of conventional columns with a limited loss in efficiency.

Since microbore columns operate at such low volumetric flow rates, they can be readily operated in the high speed mode with conventional instrumentation. To generate the similar high speed separations with conventional columns would require flow rates of 15-20 mL/min and would require modification of the pump. Figure 3 [20] is a high speed separation of 7 catecholamines in less than 1 min.

Solvent and Packing Economy

Since the volumetric flow rate is proportional to the square of the column diameter, a 95% reduction in solvent consumption can be achieved by reducing the column diameter from 4.6 mm to 1.0 mm. Scott [12] has determined the relative savings for a variety of column diameters; see Table 1. Some of the first reviews [10-13] on microbore LC stated that this savings in solvent was one of the most important advantages of microbore LC, or the most important. Analysis costs, however, are labor intensive rather than materials intensive. Thus the significant solvent savings of small bore columns are trivial when compared to the entire analysis cost. The major-

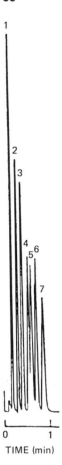

Figure 3 High speed separation of catecholamines and interfering compounds. Column: ODS-Hypersil (3 μm), 50 × 0.34 mm id. Mobile phase: MeOH-60mM potassium phosphate (pH 3.0) (8:92, v/v) containing 0.4 mM sodium *l*-octanesulphonate and 0.2 mM EDTA. Flow rate 32 μl/min; UV 220 nm. (1) 3,4-dihydroxyphenyl glycol, (2) norepinephrine, (3) epinephrine, (4) 3,4-dihydroxybenzylamine, (5) normetanephrine, (6) 3,4-dihydroxyphenylacetic acid, (7) dopamine (20 ng of each). (Reproduced with permission from Ref. 17.)

ity of companies, therefore, have not felt that the savings in solvent justifies the large expense of instrumental modification.

The savings in solvents are more significant in a research environment, where exotic solvents might be required. Lochmuller et al. [21] applied microbore HPLC to the analysis of lichen metabolites. They were able to get structural information on the perlatolic acid fraction by collecting it as it was eluted from the column with deuterated solvent and running a proton FT-NMR analysis on the fraction.

If one packs one's own columns, the savings in terms of silica gel is quite substantial as well, since a 1 mm × 25 cm column requires about 150 mg of material, compared to about 2.2 g for a 4.6 mm × 25 cm column. In terms of dollars/column, this corresponds to $3 and $44 for 1.0 mm and 4.6 mm columns respectively based on $20/g of silica gel. However, there is a limited savings when purchasing commercial columns, since a large fraction of the cost is due to quality control procedures. Again, the cost savings of microbore HPLC is most significant in an academic or industrial research laboratory, where specialized phases and solvents need to be investigated.

Rapid Thermal Equilibrium

Since small bore columns have such a high column surface-to-volume ratio, they possess excellent thermostating properties. Thus for very critical thermodynamic measurements, the heat generated by the flow of solvent through the packed bed can be readily dissipated from the center of the column to the outer column wall. The temperature gradient that develops from the column head to its end and from the column center to the walls can be minimized by reducing the column diameter.

New Detectors

Microbore columns have allowed the interfacing of HPLC to many so-called new detectors: the mass spectrometer [22–24], Fourier transform infrared [25–27], etc. The use of LC with MS has given the analyst orders of magnitude more information than a separation. The development of soft ionization techniques, such as fast atom bombardment (FAB) [28], for LC/MS will allow the analysis of molecules having a molecular weight of several thousand daltons. By coupling the LC system to detectors that also reveal structural characteristics of the compounds, the experimenter will have a tremendous amount of information at his avail.

Multidimensional Chromatography

In multidimensional separations the transfer of the entire zone of interest from one system to the next is always desired. The two main methods of zone transfer are (a) on-column concentration and (b) heart cutting. A recent paper on multidimensional chromatography [29] suggests an alternative approach: column-diameter scaling. By coupling a 1 mm system to a 4.6 mm system the authors were able to transfer the entire band of nucleosides from the first column to the second column (see Figure 4)

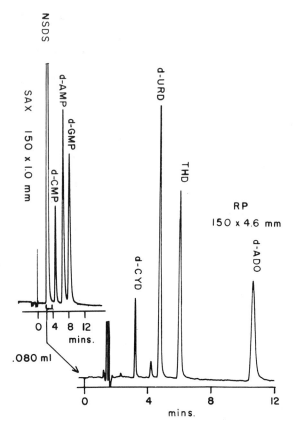

Figure 4 Multidimensional separation of the major deoxyribonucleosides and their 5'-monophosphate deoxynucleotides. Columns 25 cm × 1.0 mm; SAX column interfaced with a 15 cm × 4.6 mm RP column; mobile phases: (SAX) 0.025 M KH_2PO_4 at pH 3.9 and (RP) 90:10 (v/v) 0.025 M KH_2PO_4 at pH 3.9/methanol; UV 254. Peaks: (a) nucleosides, (b) d-CMP, (c) d-AMP, (d) d-GMP, (e) d-CYD, (f) d-URD, (g) THD, (h) d-ADO. (Reproduced with permission from Ref. 26.)

without the lengthy analysis time due to on-column concentration techniques or the inherent sampling discrimination that is associated with heart cutting.

APPLICATIONS

Neurotransmitters

Recently Wages et al. [30] have employed microbore chromatography in the monitoring of the neurotransmitter dopamine in extracellular fluid from living animal brains. They employed microbore columns for two reasons, increased mass sensitivity and decreased sample size. Using conventional columns they were not able to detect dopamine in the brain fluids to the high detection limit (5 pg), but with the 1 mm columns they were able to detect 300 fg. The other major problem with conventional columns is the sampling size. In metabolic studies the sample size should be relatively small so as not to disturb the equilibrium within the tissue of interest. Also, when sampling for a conventional column there is a increased chance of tissue damage due to the larger volumes.

A microdialysis cannula measuring approximately 300 microns in diameter was used together with a set of push-pull syringes to extract the extracellular fluid. By operating the LC at a high flow rate they were able to monitor the neurotransmitter dopamine and its metabolites. Figure 5 shows a series of chromatograms from the anterior striatum of an anesthetized rat.

Antibiotics

Antibiotics represent a chemically diverse and important group of bioactive compounds. Fielder et al. [31] employed narrow bore columns (2 mm) in the analysis of the antibiotics phosphinothricin (PTC) and PTC-Ala-Ala. Since they were interested in monitoring the two compounds during the fermentative production period, they required a system that had selectivity, sensitivity, and speed. Using precolumn derivatization they were able to detect PTC and PTC-Ala-Ala at 10 and 25 pmol respectively in less than 6 min. Figure 6 shows a chromatogram of a culture filtrate of *S. viridochromogenes* 1.6 mmol/L in PTC-Ala-Ala. Similarly, White and Laufer [32] employed 1 mm columns in the separation of cephalosporin antibiotics; see Figure 7. They noted that the microbore system has an increase in signal of 16 fold over a conventional system, which was slightly less than the theoretical value of 21.

Endogenous Compounds

One of the more sensitive detection methods for biological substances is chemiluminescence, which is similar to the light reaction of the firefly. Miyaguchi et al. [33] have applied this detection method to microbore HPLC (1 mm columns) with excellent results. Their system had detection limits of 160, 220, 220, and 280 attomol (10^{-18}/mol) for the dansylated amino acids Ala, Val, Ile, and Phe. Figure 8 shows the separation and detection of 3 fmol of each Dns amino acid.

Nucleosides, nucleotides, and nucleobases have been an area of active research for many years due to their biological importance. A recent paper by Rokushika et al. [34] employed a low capacity anion exchange resin and microbore HPLC to the separation of nucleic acid constituents. They were able successfully to separate over ten isomers. Figure 9 shows the separation of four AMP isomers. Detection limits were in the 2–5 ng range. The separation of several nucleobases, nucleosides, and nucleotides on the low capacity anion exchange capillary column is shown in Figure 10.

Other biologically active compounds of interest are the steroids. Novotny et al. [35] have used highly efficient microbore columns (>100,000 theoretical plates) for

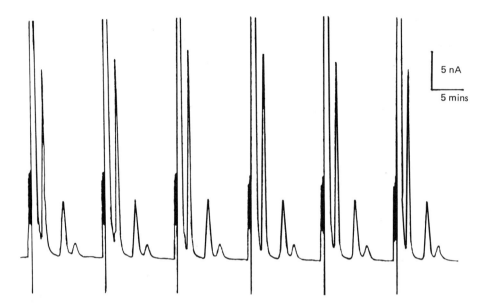

Figure 5 Series of chromatograms of dialyzed perfusate from the anterior stratum of an anesthetized rat. Sampling of the extracellular fluid is continuous. The sample loop is filled as the previous sample is chromatographed. The elution order is 3,4-dihydroxyphenylacetic acid, 5-hydroxyindole-3-acetic acid, and homovanillic acid. (Reproduced with permission from Ref. 27.)

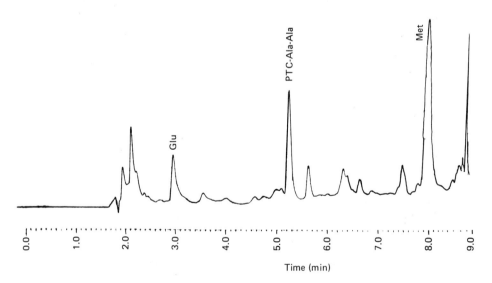

Figure 6 Separation of antibiotics from a typical biological sample. Determination of PTC-Ala-Ala in the culture filtrate of *S. viridochromogenes*. (Reproduced with permission from Ref. 28.)

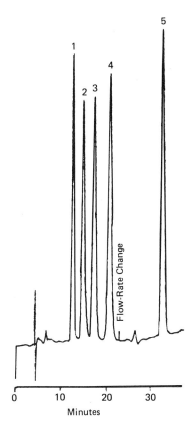

Figure 7 Separation of a mixture of cephalosporin antibiotics on a microbore column. Column: 25 cm × 1.0 mm id. μ-Bondapak C_{18} (10 μm). Mobile phase: 0.01 M sodium dihydrogen phosphate-methanol (75:25); flow rate 50 μl/min (initially), 150 μl/min (after 23 min); sample volume 5 μl. Peaks: (1) Cephalexin 0.05 μg, (2) Cefoxitin 0.05 μg, (3) Cephradine 0.07 μg, (4) Cephaloglycin 0.10 μg, (5) Cephalothin 0.23 μg. (Reproduced with permission from Ref. 29.)

the separation of urinary steroids. Figure 11 shows the chromatographic metabolic profiles of urinary steroids from a normal and a diabetic human male. Detection was by UV with an MDL of 500 pg. Sensitivity of the technique was further improved by employing a novel fluorescent reagent giving low femtogram detection limits [36]. Figure 12 shows a separation of some plasma steroids at the picogram level.

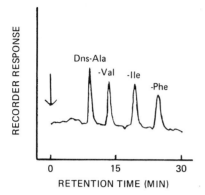

Figure 8 Chromatogram of Dns-amino acids. Detection is by chemiluminescence. Mobile phase: 0.1 M imidazole buffer (pH 7.0, NO_3)-acetonitrile (7:3, v/v); flow rate 0.03 mL/min. Reagent: 1 mM TCPO (ethyl acetate); 0.1 M H_2O_2 (acetone) (1:3, v/v); flow rate 0.6 mL/min. 3 fmol of each amino acid injected. (Reproduced with permission from Ref. 30.)

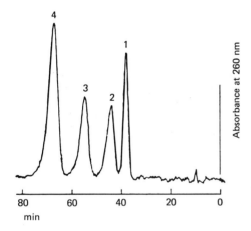

Figure 9 Chromatogram of four AMP isomers. Eluent 20 mM phosphate buffer (pH = 7.25), flow rate 1.8 μl/min, splitting ratio 120:1, detection 260 nm, sample volume 0.05 μl. Peaks: (1) 5'-AMP, (2) 3',5'-cAMP, (3) 2'-AMP, (4) 3'-AMP. Sample amount; 50 pmol of each compound.

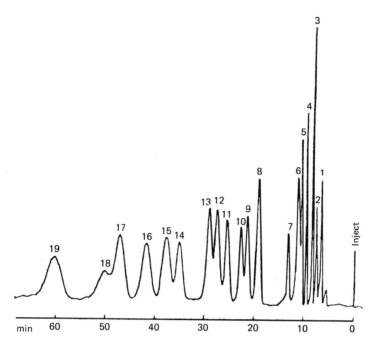

Figure 10 Chromatogram of nucleobases, nucleosides, and nucleotides. The column was preequilibrated with 17 mM phosphate buffer (pH = 7.15). Eluent 30 mM phosphate buffer (pH = 7.15) flow rate, 1.43 µl/min, splitting ratio 12o:1, detection 260 nm. Peaks: (1) Cyd, (2) Urd, (3) Cyt, (4) Ura, (5) Guo, (6) Ado, (7) Gua, (8) Ade, (9) 5'-CMP, (10) 5'-UMP, (11) 2'-CMP, (12) 3'-CMP + 2'-UMP, (13) 3'-UMP, (14) 5'-GMP, (15) 5'-AMP, (16) 2'-GMP, (17) 2'-AMP, (18) 3'-GMP, (19) 3'-AMP. Sample amount: peaks 1-8, 21.5 pmol of each compound, peaks 9-19, 26 pmol of each compound.

SUMMARY

Though microbore HPLC was initially thought to have a high potential in the pharmaceutical industry, these aspirations have not been met due to the relative inability of directly connecting them to most modern autosamplers. However the limited sample size required by these columns and their increased mass sensitivity readily lend them to the analysis of both drug degradation products and drug metabolism studies. The obvious economy of these smaller columns make them very appealing to the research community. (See also Chapter 3 by Dr. Chow for the application of microbore LC in column switching.)

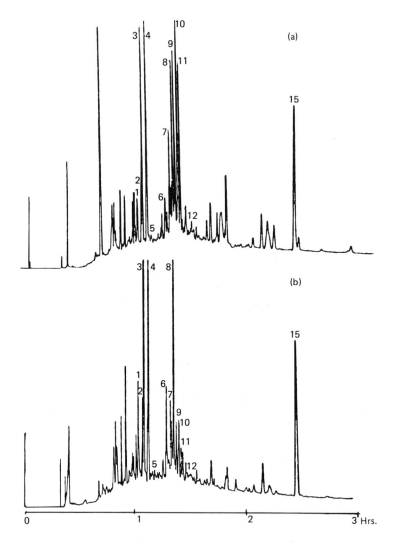

Figure 11 Comparison of representative chromatographic metabolic profiles of urinary steroids from (a) a normal and (b) a diabetic human male. Conditions: stepwise gradient 80% acetonitrile (ACN)/H_2O (15 min); 85% ACN/H_2O (14 min); 90% ACN/H_2O (15 min); 95% ACN/H_2O (18 min); 100% ACN. Column: 1 m × 0.24 mm id. packed with 3 µm spherical particles. Key: (1) 11-hydroxyandrosterone, (2) 11-hydroxyetiocholanolone, (3) allotetrahydrocortisol, (4) tetrahydrocortisol, (5) tetrahydrocortisone, (6) beta-cortolone (7) beta-cortol, (8) alpha-cortolone, (9) alpha-cortol, (10) etiocholanolone, (11) androsterone, (12) dihydroepiandrosterone, (IS) internal standard.

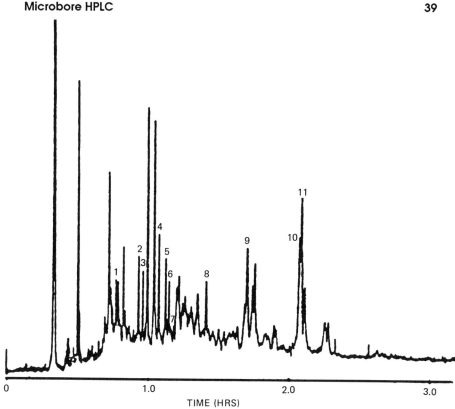

Figure 12 Chromatogram and picogram level detection of solvolysable plasma steroids on a 2.2 m slurry packed microcolumn (5 μm ODS).

REFERENCES

1. R. P. W. Scott and P. Kucera, *J. Chromatogr.*, *125*: 251 (1976).
2. C. Horvath and S. R. Lipsky, *Anal. Chem.*, *39*: 1422 (1967).
3. P. R. Brown, *High Pressure Liquid Chromatography: Biochemical and Biomedical Applications*, Academic Press, New York (1973).
4. L. R. Snyder, *J. Chromatogr.*, *63*: 15 (1971).
5. J. J. Kirkland, *J. Chromatogr. Sci.*, *9*: 206 (1971).
6. Scott and Kucera 750,000 plates.
7. R. P. W. Scott, P. Kucera, and M. Munroe, *J. Chromatogr.*, *186*: 475 (1979).
8. R. A. Hartwick and D. D. Dezaro, *Microcolumn High Performance Liquid Chromatography* (P. Kucera, ed.), *J. Chromatogr. Library*, Vol. 28 (1984).
9. N. Sagliano, Jr., S. H. Hsu, T. R. Floyd, T. V. Raglione, and R. A. Hartwick, *J. Chromatogr. Sci.*, *23*: 238 (1985).

10. R. P. W. Scott, *Small Bore Liquid Chromatography Columns: Their Properties and Uses* (R. P. W. Scott, ed.), p. 1 (1984).
11. M. Novotny, *Anal. Chem.*, *53*: 1294A (1981).
12. R. P. W. Scott, *J. Chromatogr. Sci.*, *18*: 49 (1980).
13. C. E. Reese and R. P. W. Scott, *J. Chromatogr. Sci.*, *18*: 479 (1980).
14. P. Kucera and R. A. Hartwick, *J. Chromatogr. Library*, Vol. 28: 179 (1984).
15. T. J. Filipi, *Small Bore Liquid Chromatography Columns: Their Properties and Uses* (R. P. N. Scott, ed.), p. 231 (1984).
16. H. Alborn and G. Stenhagen, *J. Chromatogr.*, *323*: 47 (1985).
17. R. P. W. Scott, *J. Chromatogr. Sci.*, *23*: 233 (1985).
18. T. Takeuchi, M. Yamazaki, and D. Ishii, *J. Chromatogr.*, *295*: 333–339 (1984).
19. I. Halasz and G. Maldener, *Anal. Chem.*, *55*: 1842 (1983).
20. D. Ishii, M. Goto, and T. Takeuchi, *J. Pharm. Biomed. Anal.*, *2*: 223–231 (1984).
21. C. H. Lochmuller, W. B. Hill, Jr., R. M. Porter, H. H. Hangac, C. F. Culberson, and R. R. Ryall, *J. Chromatogr. Sci.*, *21*: 70 (1983).
22. T. R. Covey, J. B. Crowther, E. A. Dewey, and J. D. Henion, *Anal. Chem.*, *57*: 474 (1985).
23. C. M. Whitehouse, R. N. Dreyer, M. Yamashite, and J. B. Fenn, *Anal. Chem.*, *57*: 675 (1985).
24. C. R. Blakely and M. L. Vestal, *Anal. Chem.*, *55*: 750 (1983).
25. C. M. Conroy, P. R. Griffiths, and K. Jinno, *Anal. Chem.*, *55*: 436 (1983).
26. C. C. Johnson and L. T. Taylor, *Anal. Chem.*, *55*: 436 (1983).
27. R. S. Brown, P. G. Amateis, and L. T. Taylor, *Chromatographia*, *18*: 396 (1984).
28. Y. Ito, T. Takeuchi, D. Ishii, and M. Goto, *J. Chromatogr.*, *346*: 161 (1985).
29. T. V. Raglione, N. Sagliano, Jr., T. R. Floyd, and R. A. Hartwick, *LC–GC Mag.*, *4*: 328 (1986).
30. S. A. Wages, W. H. Church, and J. B. Justice, Jr., *Anal. Chem.*, *58*: 1649–1656 (1986).
31. H. P. Fiedler, A. Plaga, and R. Schuster, *J. Chromatogr.*, *353*: 201 (1986).
32. E. R. White and D. N. Laufer, *J. Chromatogr.*, *290*: 187 (1984).
33. K. Miyaguchi, K. Honda, and K. Imai, *J. Chromatogr.*, *316*: 501 (1984).
34. S. Rokushika, Z. Y. Qiu, and H. Hatano, *J. Chromatogr.*, *320*: 335 (1985).
35. M. Novotny, M. Alasandro, and M. Konishi, *Anal. Chem.*, *55*: 2375–2377 (1983).
36. M. Novotny, *J. Pharm. Biomed. Anal.*, *2*: 207 (1984).

3

Column Switching Techniques in Pharmaceutical Analysis

Francis K. Chow

Par Pharmaceutical, Inc., Spring Valley, New York

INTRODUCTION

In pharmaceutical analysis, assay specificity is always an important concern. A specific HPLC assay has to be able to resolve the analyte from its potential interferences. For drug substances assay, resolution from its synthetic impurities and potential degradation products has to be demonstrated. In analyzing pharmaceutical formulations and biological samples, resolution from the sample matrices has also to be considered. This aspect of assay specificity is particularly emphasized in recent FDA publications on the requirements in NDA submission [1]. With the advances in HPLC column technology and the better understanding of the separation mechanisms, for many common analytical problems, high performance chromatographic methods can be developed relatively easily to achieve the required separations. However, when the samples are more complex, off-line sample preparation becomes necessary to isolate and concentrate the analytes from their sample matrices prior to analysis. These manual off-line sample preparations are quite time-consuming, and they limit sample throughput. To accommodate the large number of samples commonly encountered in many pharmaceutical laboratories, automation of sample preparation becomes highly desirable.

Typical sample preparations are primarily liquid-liquid, liquid-solid (sorbent) extractions, solvent evaporations, and sample reconstitutions with suitable solvents for HPLC analysis. With the advances in microprocessor controlled HPLC instrumentation and the availability of high precision, low dead volume switching valves, columns can be coupled in any configuration rather easily to automate sample preparations and enhance system resolution capability. This approach of automated sample preparation using column switching techniques is growing in popularity in the analysis of biological samples and complex pharmaceutical formulations. A survey of the literature indicates that more than 100 papers have been published in this area in the past five years.

Column switching is a general term; it is also referred to as multidimension, multiphase, multicolumn, or coupled column chromatography. In principle, this technique involves the separation of a complex mixture by selectively transferring a portion of a separation zone from one column to another for further separation. Switching valves with microprocessor controlled actuators are commonly used to automate the switching and zone transferring. The idea of using column switching to enhance system selectivity and automate sample preparation is not new. It was first introduced by Snyder in 1964 [2,3] for low pressure LC separation and later in HPLC application [4,5]. In 1973, Huber et al. discussed the general principle of column switching in HPLC [6]. Recently, this topic has been reviewed by Freeman [7], Majors [8], and Raglione et al. [9]. This chapter will discuss some theoretical considerations and practical applications of the technique.

BACKGROUND CONSIDERATIONS

Effect on Peak Capacity

The ultimate goal for any chromatographic separation scheme is to achieve the maximum sample information in the shortest possible time. In other words, optimization of system resolution is a primary concern. The resolution Rs, in general, is controlled by three parameters: system efficiency, usually expressed in the number N of theoretical plates; system selectivity α; and capacity ratios k', which determine the retention times. This relationship is given by

$$\text{Rs} = \frac{1}{4}(\alpha - 1) \sqrt{N} \frac{k'}{1 + k'} \qquad (1)$$

Much research effort in the past decades has been focused on the optimizations of these three parameters in many separation systems. The rapid advances in column technology have produced ultra-high performance columns with efficiency approaching the theoretical limit. There is little room for improvement in the parameter N. Similarly, the continuing understanding in separation mechanisms, α and k' can be easily manipulated by the use of novel bonded phases or unique mobile phase additives. These research efforts have pretty much exhausted the maximum resolution power that a single column can produce. This leads to an important concept of peak capacity and how to extend it through coupled columns. In a single column system, no matter how one manipulates the separation by optimizing the three parameters described above, there is always an inherent limitation on the number of peaks one can fit into a given chromatographic space. This concept is extensively discussed by Giddings [10–12]. The relationship of poor capacity, system efficiency and capacity ratio is given by:

$$\Phi = 1 + \frac{\sqrt{N}}{m} \ln(1 + k') \qquad (2)$$

where m is the number of standard deviations σ that are taken as equal to the peak width [12] and k' is the capacity ratio of the last eluting peak. A typical high performance LC system has a peak capacity in the range of 30 to 150. When the sample complexity increases and exceeds the system's peak capacity limit, chromatographic peaks will start overlapping. However, the system's peak capacity limit can be extended by column coupling. As discussed by Freeman [7] and Raglione et al. [9], the increase of peak capacity when several columns are coupled lies between two limits: the lower limit is reached when all columns coupled are redundant and the upper limit when all columns are nonredundant. When nonredundant columns are coupled, each having a peak capacity Φ_i, there is a multiplicative effect on the increase of total peak capacity Φ_T as given by

$$\Phi_T = \prod_{i=1}^{n} \Phi_i \qquad (3)$$

Moreover, if each column has the same peak capacity, Equation 3 will be reduced to the simple exponential given by

$$\Phi_T = \Phi^n \qquad (4)$$

In the case of coupling redundant columns, it can be considered as increasing column length and consequently system efficiency. Assuming n identical redundant columns are coupled, Equation 2 can be written as

$$\Phi_T = 1 + \frac{\sqrt{(nN)}}{m} \ln(1 + K'n) \qquad (5)$$

or

$$\Phi_T \cong n^{\frac{1}{2}} \Phi \qquad (6)$$

Equation 6 reveals a fact that is well known to chromatographers, that a 100 fold increase in column length is needed to decrease 10 fold in relative peak width. These effects of increasing peak capacity in coupling redundant and nonredundant columns are illustrated in Figure 1, and some examples of the effect of redundancy on peak capacity are given in Table 1 [9].

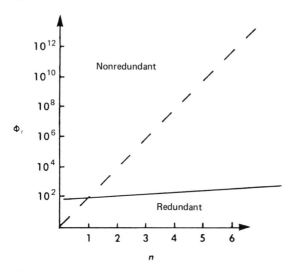

Figure 1 The effect of system redundancy on total peak capacity for n coupled columns plotted assuming equal peak capacities for each column. (From Ref. 9, used with permission.)

Theoretically, it appears that coupled columns allow exponential enhancement of both a system's selectivity and its efficiency. In practice, however, when several columns are connected, this gain can be easily offset by excessive losses of system efficiency due to the additional extracolumn effects in connecting columns and the difficulty in transferring separation zones from one column to another, particularly when mobile phases are not compatible. This issue of band broadening in coupled columns will be discussed below.

Effect on Band Broadening

When an analyte is injected into a chromatographic system, the chromatographic zone will start to spread. The extent of this zone spreading depends on the diffusion kinetics within the analytical column and the contributions due to extracolumn ef-

Table 1 Effect of Redundancy on Peak Capacity[a]

Modes	Φ_1	Φ_2	Φ_T
SEC-RPC[b]	30	100	3000
IEC-RPC[c]	70	100	7000
LC-GC	70	300	2100
IEC-IEC	70	70	99
RPC-RPC	100	100	141

fects. In the past decades, band broadening has been a topic of extensive research [13–15]. When several columns are coupled, the effect on band broadening can be expressed mathematically by equation 7, as discussed by Tomellini et al. [16] and Dolphin et al. [17]

$$\sigma_2^t = \sum_{i=1}^{n} \frac{L_i^2}{N_i} + \sigma_{ec}^2 \qquad (7)$$

where σ_t^2 is the total band variance, σ_{ec}^2 is the extracolumn variance due to the injection system, switching valves, connecting tubings, couplings, detectors, etc., and L_i and N_i are column length and column efficiency respectively. Huber et al. experimentally determined the variances due to the switching valve and concluded that its contribution to the overall band broadening was significant in comparison with the rest of the system [6]. The connecting tubing, particularly when used as bypass coil [17], is one of the most significant contributions to band broadening. In designing any column switching configuration, the equation developed by Scott et al. [18] for band variance contributed by connecting tubings (σ_b^2) should be considered:

$$\sigma_b^2 = \frac{\pi r^4 l}{24 D_m Q} \qquad (8)$$

where r is the internal diameter of the tubings, l is the tubing length, Dm is the diffusivity of the solute in the mobile phase, and Q is the flow rate of the mobile phase. Obviously, by reducing r and l and increasing Q, the extracolumn effect due to connecting tubing would be minimized.

Another important source of band broadening is the injection volume as discussed by Karger et al. [19] and Martin et al. [20]. For a given injection technique and sample component, the contribution of injection volume (Vs), the band broadening (σ_i^2) is given by

$$\sigma_i^2 = \frac{V_s^2}{K^2} \qquad (9)$$

where K is a constant. In a single column system, the injection volume can be easily adjusted to limit its contribution to the overall band broadening. In coupled column systems, however, the injection volume of the secondary column is determined by the transfer volume of the primary column. This transfer volume can be quite large, causing unacceptable dispersion. In practice, this transfer volume can be effectively reduced by a peak compression technique [21] in which the solute can be concentrated on the head of the secondary column by introducing a weaker eluting solvent.

In transferring separating zones, Snyder [22] has defined some empirical rules to minimize band broadening: (a) use columns with similar internal diameters; (b) use

Table 2 Types of HPLC that Are Easily Coupled On-Line

	Primary mode					
	LSC	N-BPC	RPC	IEC	GPC	GFC
LSC	•	•			•	
N-BPC	•	•			•	
RPC			•	•	•	•
IEC			•	•	•	•
GPC	•	•	•	•	•	
GFC			•	•		•

LSC = liquid-solid (adsorption) chromotography (such as silica or alumina phases).
N-BPC = normal bonded phase chromatography (such as amino, cyano, or diol phases).
RPC = reversed phase chromatography
IEC = ion exchange chromatography
GPC = gel permeation chromatography (size exclusion chromatography using nonaqueous mobile phases
GFC = gel filtration chromatography (size exclusion chromatography using aqueous mobile phases
Source: From Ref. 8, used with permisssion.

columns with similar efficiency; (c) keep k' at optimum range for all columns. However, when the mobile phase of the primary column is not compatible with the secondary column, it will cause excessive loss of resolution. For example, the transfer of a large volume of organic rich mobile phase in a separation zone from a size exclusion column to a reversed phase column will cause a partial migration of the zone and decrease resolution [23]. To overcome this problem, two general approaches have been used: (a) a peak compression technique as described above; (b) cutting of the transfer volume to minimize zone spreading. Both methods have some limitations. For example, peak compression requires additional plumbing to introduce a weaker, mutually miscible solvent in between the two columns as reported by Linder et al. [24]. The peak cutting technique suffers loss of sensitivity, since only a small portion of the sample is transferred. Reproducibility is also reported to be quite poor [25,26]. An attractive alternative is to couple a microbore column to a common analytical column so that total zone transfer can be accomplished without excessive zone spreading [9]. Table 2 summarizes the different modes of HPLC that can easily be coupled [8].

BASIC INSTRUMENTATION

A generalized setup of multidimensional HPLC is shown in Figure 2. As discussed in the previous section, mobile phase compatibility of different modes of columns, and the additional extracolumn effects introduced by the switching valves, tubings, size of transfer volumes, etc., will determine the optimum configuration for a particular application. In most cases, a much simpler setup than in Figure 2 will be ade-

Column Switching Techniques

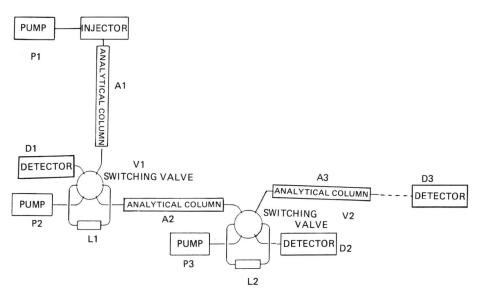

Figure 2 Generalized setup of column switching.

	Columns					Valves		Pumps			Detectors		
	A1	A2	A3	L1	L2	V1	V2	P1	P2	P3	D1	D2	D3
Applications													
Trace enrichment	N	Y	O	Y	O	Y	O	Y	Y	O	N	O	Y
Sample cleanup	O	Y	O	O	O	Y	O	Y	Y	O	O	O	Y
Fraction cutting	Y	Y	O	N	O	Y	O	Y	Y	O	O	O	Y

N= no; Y = yes; O = optional. Operations in V1 can be repeated in V2, V3, etc., if necessary; therefore in all cases A3, L2, V2, and P3 are listed as optional.

quate to do the job. The setup in Figure 2 basically reveals two important modes of operations commonly used in column switching: automated sample preparation and enhancement of system resolution. The small columns, L1, L2, etc., located in the switching valve loop are commonly used for automated sample preparation such as trace enrichment, peak compression, sample cleanup, etc.; since they are located on the valve loop, they are sometimes referred to as loop columns. Typical length of these columns are 3 to 5 cm with similar internal diameter to the analytical columns. Depending on the applications, packings of these columns can be different from the analytical columns. The analytical columns, A1, A2, etc., however, are connected in series to provide multidimensional separation. When heart cutting or front cutting techniques are used in transferring separation zones from one column to another to minimize zone spreading, the loop column normally will be replaced by an injection loop, and the size of the loop will determine the size of the transfer volume. Switching valves are normally controlled by pneumatic/electrical actuators interfaced to the system microprocessor that orchestrates the switching sequences. Extra care has to be exercised to minimize unnecessary extracolumn effects.

Columns of different separation modes can also be coupled, as shown in Figure 3. This configuration of column switching is very useful in automated method development. The chromatographic behavior of a sample mixture can be sequentially screened unattendedly with different mobile phases on different columns controlled by the system microprocessor.

Column switching accessories are readily available from various manufacturers. Compact self-contained column switching devices and built-in units are also marketed by several manufacturers. Table 3 summarizes various commercial sources. Almost any column switching configuration can be built and tailored to each ana-

Table 3 Commercially Available Column Switching Devices and Accessories

Items	Manufacturers
(A) Acessories	
Switching valves	Rheodyne, Cotati, Calif.
Actutuators	Valco, Houston, Tex.,
Solenoid interface	Autochrome, Milford, Mass.
Solenoid power supply	
Loop column cartridges/holders	Brownlee Lab., Santa Clara, Calif.
	EM Science, Cherry Hill, N. J.
(B) Devices	
Waters automated valve station	Millipore, Waters, Milford, Mass.
Mcs 670 Kontron Tracor column switching unit	Kontron, Zürich, Switzerland
Model 410 column switching system	Autochrom, Milford, Mass.
Model 460 column selector	
Tandem model 7067 switching device	Rheodyne, Cotati, Calif.

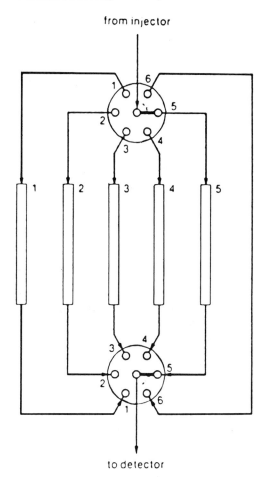

Figure 3 Typical column switching setup for automated method development.

lytical problem with the commercially available accessories. However, compact and self-contained column switching units are quite handy and easy to use.

APPLICATIONS

Pharmaceutical Formulations

Many pharmaceutical formulations such as creams, ointments, injectables, syrups, suppositories, and transdermal patches contain complex matrices. Analysis of the active components in these formulations, in many cases, requires substantial sample preparation such as liquid-liquid partition, dissolution, filtration, open column chromatography, minicolumn and cartridges, evaporation and reconstitution of sample residue [27]. These manual sample preparations are time-consuming and limit sample throughput. Obviously, it is highly desirable to automate the sample preparation. Column switching offers a great opportunity to automate these manual procedures. Kenley et al. reported the use of a microprocessor controlled, tandem reversed phase HPLC method for analyzing both cream and ointment formulations [28-30]. Creams and ointments are complex pharmaceutical preparations containing many components such as active drugs, preservatives, antioxidants, cream and ointment bases, etc. Sample cleanup is quite time-consuming [31,32]. In Kenley's approach, precolumn clean-up consisted of dissolving samples in a THF and isopropanol mixture, centrifuging and removing sediments, and directly injecting into the column switching system. The switching valves were operated in a front cutting mode directly only the analytes and internal standard onto the analytical column, while the retained excipient materials were black-flushed to waste. This method showed excellent accuracy and precision and reduced sample preparation times by a factor of nearly 3. A typical chromatogram for the assay of a cream formulation is shown in Figure 4. It should be noted that a short loop column is adequate for this front cutting operation so that the separation zone can be transferred rapidly onto the analytical column without excessive dispersion and the retained materials can be easily black-flushed to waste. A similar study was also reported by E. Miller et al. [33] for the analysis of topical creams using on-line sample cleanup, shown in Figure 5. A Waters automated valve station was used to facilitate the column switching operation. A short Waters Guard-PAK precolumn module and inserts were used for sample cleanup.

Other pharmaceutical formulations that require tedious sample cleanup are cough syrups. Cough syrups usually are multicomponent mixtures containing antitussive and decongestant or expectorant agents formulated with liquid sugar, sorbitol, dyes, flavoring agents, and preservatives. For assaying the stability of the active ingredients in the formulation, it is desirable to remove as much excipient interference as possible from the sample so that the active components and their degradation products are the primary occupants of the limited chromatographic "space." A procedure was developed in the author's laboratory for the analysis of active ingredients in syrup formulations [34]. As shown in Figure 6, a 3 cm cation exchange column L1 was used to retain the active drug substances in the cationic forms while the excipients including liquid sugar, sorbitol, parabens, etc. were washed to waste. The valve was then switched to black-flush the actives onto a reversed phase analytical column for analysis. Typical chromatograms are shown in Figure 7. The cation exchange loop column used in this work not only cleaned up the sample but also concentrated the analytes, thus reducing dispersion of the transfer volume. Since only the drugs of interest are retained and chromatographed by the reversed phase column excellent specifically was achieved. It should be noted that water of the highest purity should be used in any column switching system to minimize interferences. In this work, the size of the solvent peak was substantially reduced when HPLC grade water is used.

Another application in the author's laboratory [35] is to use the column switching technique in the trace enrichment mode. A large volume of a neonatal injectable containing a very much diluted solution of Naloxone was injected directly onto a 3 cm reversed phase column with a mobile phase containing a low percentage of or-

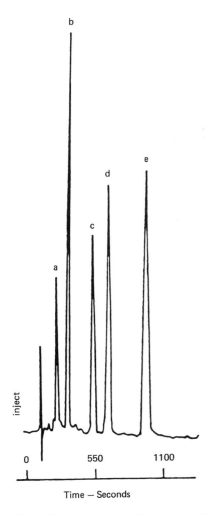

Figure 4 Chromatogram for the assay of a developmental corticosteroid cream formulation using column switching technique. Peaks a to c are degradation products, peak d is internal standard, and peak e is the drug, (From Ref. 28, used with permission.)

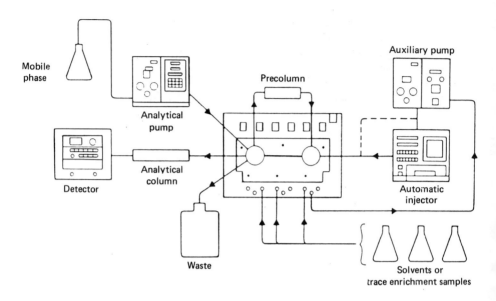

Figure 5 HPLC system configuration for on-line cleanup and analysis of topical cream using Waters valve station. (From Ref. 33, used with permission.)

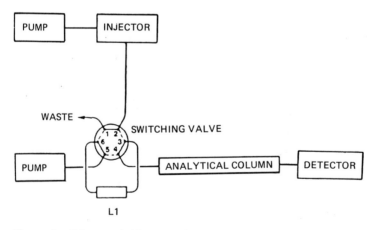

Figure 6 Column switching setup for automated assay of cough syrup; L1 is a 3 × 4.6 mm id cation exchange column; analytical column is a 25 cm × 4.6 mm id C8 reversed phase column.

Column Switching Techniques

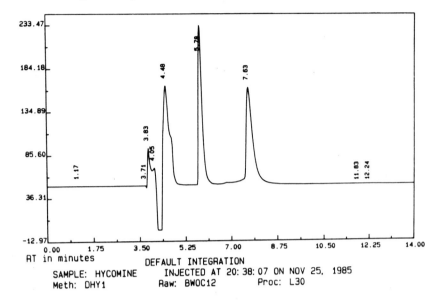

Figure 7 Chromatogram for the assay of a syrup formulation using column switching technique.

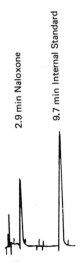

Figure 8 Chromatogram for the assay of a low concentration injectable using column switching trace enrichment technique. 1 mL of sample is injected.

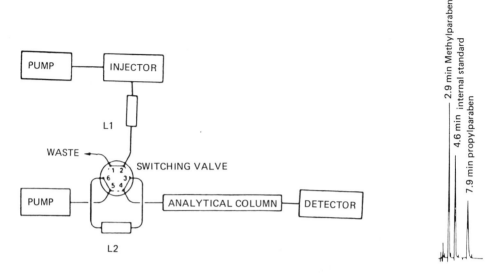

Figure 9 Column switching setup for a "universal" automated assay for parabens in pharmaceutical formulations. L1 is a 3 cm × 4.6 mm id cation exchange column; L2 is a 3 cm × 4.6 mm id C8 column. Analytical column is 25 cm × 4.6 mm id C8 column.

ganic solvent, to concentrate the drug on the head of the column. A stronger mobile phase was then applied by valve switching to elute the trapped analyte onto a reversed phase column to perform the separation as shown in Figure 8. Minimal band broadening is observed even as up to 1 ml of sample was injected. Samples can be processed on line unattendedly with a throughput increased by more than 200% in comparison with a liquid-liquid extraction technique. The system setup is a simplified version of Figure 2 in which A1, A3, L2, P3, D1, D2, and V2 are deleted and A2 is connected to D3.

A third interesting application in our laboratory [36] was to develop a universal automated method to the assay paraben preservatives. Many liquid formulations of drugs with amino functional groups use methylparaben and propylparaben as preservatives. Excipient complexity varies from formulation to formulation; therefore different assays for parabens are normally required for different formulations. To eliminate the matrix effect, a column switching setup is shown in Figure 9 can be used. Two loop columns were utilized. The cation exchange loop column L1 was used to trap all actives and excipients containing amino functional groups; salt buffer in injectables and liquid sugar or sorbitol in syrups were washed to waste through the reversed phase loop column L2. The parabens, however, were retained on the head of the column L2 in the organic lean loading solvent. After the loading

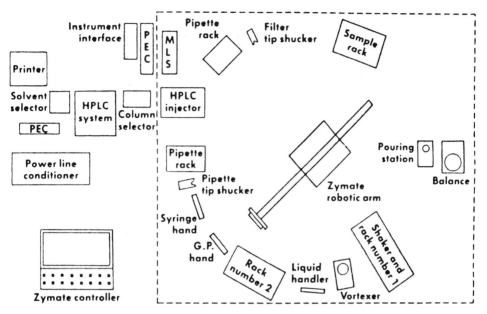

Figure 10 Diagram showing the robotic system's module and instrument arrangements. Column selector is similar to Figure 3.

sequence was completed, a stronger mobile phase was used to elute the parabens onto a reversed phase analytical column to perform the separation. Sample preparation was simple. Internal standard was pipetted into the sample and diluted accordingly so that one set of standards could be used for all samples, syrups or injectables. The system allows up to 100 injections to be made before the cation ion exchange loop column started leaching of solutes. The column was either replaced or regenerated. The regeneration procedure can also be automated on line by adding another valve and using an additional pump to deliver the regeneration solvent in an appropriate sequence.

In addition to automated sample preparation, Antwerp et al. [37] reported a highly sophisticated automation design by interfacing a Zymark robotic arm to a programmable HPLC system with a multiple solvent selector and a column switcher to enhance system flexibility and efficiency, as shown in Figure 10. The system was programmed to process different types of samples with robotic handling for analysis by the appropriate column and solvent. Consequently, this system provides the capability of unattended handling of a large number of samples, each requiring different HPLC assay. Moreover, if sample stability in solution is a concern, this system has the advantage of immediate sample dissolution before analysis, thus minimizing sample degradation.

Biological Samples

A survey of literature showed that more than 80% of column switching works were in the area of analyzing drugs in biological fluids. Recently, Roth and Edholm et al. [38,39] have reviewed the use of coupled column techniques for the pharmacological testing of drug substances, covering works to 1984.

Drug analysis in biological fluids has always been a challenge to analytical chemists due to the complexity of the sample matrix and the low levels of the drugs. For example, urine samples contain a wide variety of components including salts, aromatic acids, catecholamine metabolism, etc. [40-42]; and plasma consisted of large molecular weight proteins, lipids and fats, etc. Many components in these matrices are often not compatible with the mobile phases used in modern LC, and so direct injection of a serum or urine sample onto an HPLC system without prior sample preparation is deleterious. The classical approach of sample preparation for biological fluids commonly includes deproteinization, extraction, evaporation, and reconstitution with a suitable solvent for injection onto a column. These procedures can be automated by column switching techniques. Some examples of recent work are summarized below.

Basically, there are two types of approaches in using column switching in the assay of biological fluids: (a) preliminary sample cleanup such as deproteinization prior to analysis [43-55]. The deproteinization step will ensure that the high molecular weight proteins in the serum samples will not cause column clogging. (b) Direct injection of diluted samples for simultaneous sample cleanup, enrichment, and analysis [56-64]. Takahagi et al. [56] developed a procedure involving direct injection of plasma into a column switching system containing an aqueous gel chromatography column and an ODS column. The first column was effective in the separation of water-soluble polymers such as proteins, and the drugs of interest were then introduced into the ODS column for further separation and quantitation. The validity and applicability of this procedure were demonstrated in the analysis of the antibiotic cefmetazole, the anticoagulant warfarin, the antitumour agent carboquone, and the anaesthetic ketamine in several metabolic studies. Timm et al. [57] reported an automated HPLC assay for the determination of an aldosterone antagonist in urine by direct injection of diluted urine samples onto a C18 cleanup column to remove the polar urine components, and the retained substances were then black-flushed onto an analytical C18 column for separation and quantitation. Low detection limit and high precision were achieved. Column switching, as demonstrated in these techniques, is a viable approach in the automation of bioanalytical assay.

Miscellaneous

Column switching was also applied to preparative isolations. Little et al. reported the use of column switching in semipreparative isolation of the sweetener stevioside, an extract from the dried leaves of Stevia rebaudiana, with a Kontron "TRACER" MCS 670 column switching unit [65]. A sample was injected onto a semipreparative column in a gross overloaded condition to increase yield. The fraction of eluent containing stevioside was heart cut and transferred onto a second

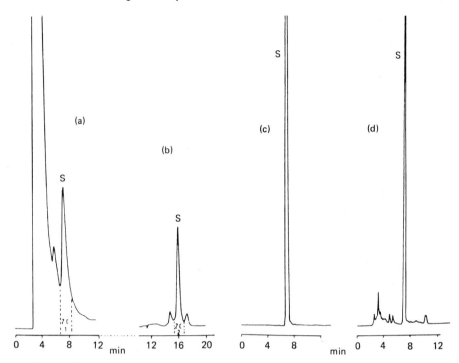

Figure 11 Chromatograms showing semipreparative isolation of stevioside (S) with timing of cutting zones (ZC). (a) First semipreparative column under grossly overloaded condition. (b) Second semipreparative column. (c) Chromatogram showing purity of the isolated stevioside. (d) Purity of conventional isolate. (From Ref. 65, used with permission.

semipreparative column, which gave improved resolution of the analyte because the column was not grossly overloaded. As shown in Figure 11, stevioside can be isolated with higher purity than by the conventional approaches.

Column switching techniques have also been reported in the assay of analytes in food and feed samples. Jaumann et al. reported the use of an on-line trace enrichment technique to isolate vitamin B_2 in food samples including instant chocolate drink, grape sugar mix, and beer [66]. The amount of vitamin was in the ppm level, and a better than 3% of reproducibility was reported. Lepom reported the use of column switching to determine sterigmatocystin in feed at low ppb level [67]. Two reversed phase columns were used to allow direct switching of an effluent segment of the precolumn to the analytical column without trapping loop or solvent evaporation steps. The column length, the alkyl chain length of the bonded phase, and the particle size of the silica were selected in such a way that column 1 could be eluted with a markedly weaker solvent than column 2 in order to minimize band spreading.

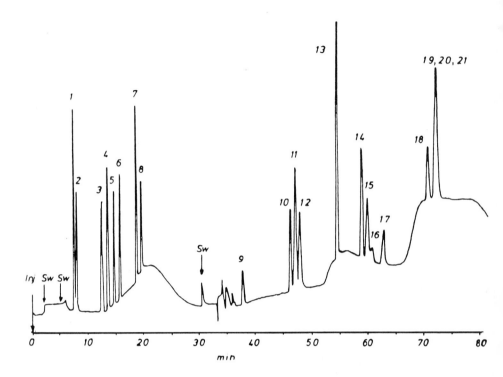

Figure 12 Two-column HPLC with ODS column followed by an anion-exchange column. Chromatogram of a synthetic mixture containing purine bases, nucleosides, and nucleotides. Peaks: 1 = hypoxanthine.

Another interesting application is in the analysis of purine nucleic acid components. Lang et al. reported a column switching setup that combines reversed phase and anion exchange chromatography for the separation of purine bases, nucleosides, and nucleotides [68]. The sample was injected onto the reversed phase column, which was initially, connected to an ion exchange column. The bases were retained on the reversed phase column, while the more polar nucleosides and nucleotides were eluted onto the ion exchange column. After the polar analytes had reached the ion exchange column, a valve switching step was then applied to divert the effluent exciting the reversed phase column to a UV detector. Chromatography on the reversed phase column was performed with a mobile phase gradient. When the reversed phase separation was completed, the anion exchange column was eluted with a phosphate buffer gradient of increasing ionic strength. The typical separation is shown in Figure 12.

Column switching techniques have also been reported for the separation of proteins [69] and hydrocarbons of different types [70].

CONCLUSION

With the availability of microprocessor controlled HPLC instrumentation, the use of column switching techniques to automate sample preparation and to enhance system selectivity and efficiency becomes preferred over many complex manual techniques. With these low-cost microprocessor controllers, almost any HPLC configuration can be tailored to meet the analytical needs in every area of pharmaceutical analysis, research and development, quality control, and process control. Undoubtedly, column switching techniques have made a major contribution to the advances in HPLC automation. HPLC column switching developments will find applications in many other areas in analytical chemistry. Development with column switching in microbore LC will be favored due to small transfer volume which will keep zone spreading to a minimum in switching from one column to another. In the area of LC MS interface, column switching might be used to exchange the chromatographic eluent for a solvent that is compatible with mass spectroscopic analyses. Research activities in these and many other areas will be evident in the future.

ACKNOWLEDGEMENT

The author wishes to thank E. I. du Pont de Nemours & Company for their support in the preparation of this manuscript during his employment with them.

REFERENCES

1. FDA Publication, Fed. Reg., Vol. 50, No. 36, 7452-7519, (1985)
2. L. R. Snyder, *Anal. Chem.*, *36*: 774 (1964).
3. L. R. Snyder, *Anal. Chem.*, *37*: 713 (1965).
4. L. R. Snyder, in *Modern Practice of Liquid Chromatography* J. J. Kirkland, ed., Wiley Interscience, New York, pp. 232-236 (1977).
5. L. R. Snyder, J. W. Dolan, and Sj. van der Wal, *J. Chromatogr.*, *203*: 3 (1981).
6. J. F. K. Huber, R. Van der Linden, E. Ecker, and M. Oreans, *J. Chromatogr.*, *83*: 267 (1973).
7. D. H. Freeman, *Anal. Chem.*, *53*: 2 (1981).
8. R. E. Majors, *LC Magazine*, *2*: 358 (1984).
9. T. V. Raglione, N. Sagliano, Jr., T. R. Floyd, and R. A. Hartwick, *LC Magazine*, *4*: 329 (1986).
10. J. C. Giddings, *Anal. Chem.*, *39*: 1027 (1967).
11. J. C. Giddings, *Anal. Chem.*, *56*: 1295A (1984).
12. J. M. Davis and J. C. Giddings, *Anal. Chem.*, *57*: 2168 (1985).
13. E. Grushka, L. R. Snyder, and J. H. Knox, *J. Chromatogr. Sci.*, *13*: 25 (1975).
14. E. Grushka, L. R. *Anal. Chem.*, *46*: 510A (1974).
15. J. C. Giddings, in *Dynamics of Chromatography*, Part 1, Marcel Deckker, New York, (1965).
16. S. A. Tomellini, S. Hsu, and R. A. Hartwick, *Anal. Chem.*, *58*: 904 (1986).
17. R. J. Dolphin and F. W. Willmott, *J. Chromatogr. Sci.*, *14*: 585 (1971).
18. R. P. W. Scott and P. Kucera, *J. Chromatogr. Sci.*, *9*: 641 (1971).
19. B. L. Karger, M. Martin, and G. Guiochon, *Anal. Chem.*, *46*: 1640 (1974).

20. M. Martin, P. L. Joynes, A. J. Davies, and J. E. Lovelock, *Anal. Chem.*, *43*: 1966 (1971).
21. M. Broquarie and P. R. Guinebault, *J. Liq. Chromatogr.*, *4*: 2039 (1981).
22. L. R. Snyder and J. J. Kirkland in *Introduction to Modern Liquid Chromatography*, 2nd edition, Wiley Interscience, New York, 696 (1979).
23. E. L. Johnson, R. Gloor, and R. E. Majors, *J. Chromatogr.*, *149*: 571 (1978).
24. W. Lindeer, H. Ruchendorfer, and W. Lechner. Paper presented at VIIth International Symposium on Col. Liquid Chromatography, Baden-Baden, Federal Republic of Germany (1983).
25. R. E. Majors, *J. Chromatogr. Sci.*, *18*: 571 (1980).
26. E. L. Colling, B. H. Burda, and P. A. Kelly, *J. Chromatogr. Sci.*, *24*: 7 (1986).
27. J. Mills, *Biochem. Soc. Trans.*, *3*: 1073 (1984).
28. D. L. Conley and E. J. Benjamin, *J. Chromatogr.*, *257*: 337 (1983).
29. E. J. Benjamin and D. L. Conley, *Int. J. Pharm.*, *13*: 205 (1983).
30. R. A. Kenley and S. Chaudry, *Drug Development and Industrial Pharmacy*, *11*: 1781 (1985).
31. P. A. Williams and E. R. Biehl, *J. Pharm Sci.*, *70*: 530 (1981).
32. M. C. Olson, *J. Pharm. Sci.*, *62*: 2001 (1973).
33. E. Hiller, R. Cotter, and M. Andrews, *Amer. Lab.* Aug. (1985).
34. F. K. Chow, unpublished data (1980).
35. F. K. Chow, unpublished data (1981).
36. F. K. Chow, unpublished data (1981).
37. J. V. Antwerp and R. F. Venteicher, LC-GC Magazine, *4*: 458 (1986).
38. W. Roth, in *LC in Pharmacological Development* (I. W. Wainer, ed.), Aster Publishers, 323 (1985).
39. L. Edholm and L. Orgen, in *LC in Pharmacological Development* (I. W. Wainer, ed.), Aster Publishers, 345. (1985).
40. C. D. Scott, *Clin. Chem.*, *14*: 521 (1968).
41. I. Molnar, Cs. Horvath, and P. Jatlow, *Chromatographia*, *11*: 260 (1978).
42. M. Spiteller and G. Spiteller, *J. Chromatogr.*, *164*: 253 (1979).
43. D. L. Reynolds and L. A. Pachla, *J. Pharm. Sci.*, *74*: 1091 (1985).
44. G. Hamilton, E. Roth, E. Wallisch, and F. Tichy, *J. Chromatogr.*, *341*: 411 (1985).
45. J. C. Jordan and B. M. Ludwig, *J. Chromatogr.*, *362*: 263 (1986).
46. J. Carlqvist and D. Westerlund, *J. Chromatogr.*, *344*: 285 (1985).
47. C. Julien, C. Rodriguez, G. Cuisnaud, N. Bernard, and J. Sassard, *J. Chromatogr.*, *344*: 51 (1985).
48. D. J. Gmur, G. C. Yee, and M. S. Kennedy, *J. Chromatogr.*, *344*: 422 (1985).
49. R. H. Pullen, and J. W. Cox, *J. Chromatogr.*, *343*: 271 (1985).
50. C. R. Benedict and M. Risk, *J. Chromatogr.*, *317*: 27 (1984).
51. G. Decristoforo, *Anal. Chim. Acta*, *163*: 25 (1984).
52. W. C. Pickett and M. B. Marietta, *Prostaglandins*, *29*: 83 (1985).
53. H. T. Smith and W. T. Robinson, *J. Chromatogr.*, *305*: 353 (1984).
54. L. E. Edholm and B. M. Kennedy, *Chromatographia*, *16*: 341 (1982).
55. J. W. Cox and R. H. Pullen, *Anal. Chem.*, *56*: 1866 (1984).
56. H. Takahagi, K. Inoue, and M. Horiguchi, *J. Chromatogr.*, *352*: 369 (1986).
57. U. Timm and A. Saner, *J. Chromatogr.*, *378*: 25 (1986).
58. J. Dows, M. Lemar, A. Frydman, and J. Gaillot, *J. Chromatogr.*, *344*: 275 (1985).
59. L. Weidolf, *J. Chromatogr.*, *343*: 85 (1985).
60. J. B. Lecaillon, C. Souppart, and F. Abadie, *Chromatographia*, *16*: 158 (1982).

61. P. O. Edlund and D. Westerlund, *J. Pharm. Biomed. Anal.*, 2: 315 (1984).
62. T. Shimizu, M. Kubo, and K. Nakagawa, *Yakugaka Zasshi, 103*: 1174 (1983).
63. O. Stetten, P. Arnold, M. Aumann, and R. Guserle, *Chromatogrphia, 19*: 415 (1984).
64. U. Juergens, *J. Chromatogr., 310*: 97 (1984).
65. C. J. Little and O. Stahel, *J. Chromatogr., 316*: 105 (1984).
66. G. Jaumann and H. Englehardt, *Chromatographia., 20*: 615 (1985).
67. P. Lepom, *J. Chromatogr., 354*: 518 (1984).
68. H. R. M. Lang and A. Razzi, *J. Chromatogr., 356*: 115 (1986).
69. H. Lindblom and L. G. Fagerstam, *LC Magazine, 3*: 360 (1985).
70. T. V. Alfredson, *J. Chromatogr., 218*: 715 (1981).

Part Two
Specialized Detection Techniques

4

Liquid Chromatography/Electrochemistry in Pharmaceutical Analysis

Peter T. Kissinger

Bioanalytical Systems, Inc., West Lafayette, Indiana

Donna M. Radzik

Lederle Laboratories Division, American Cyanamid Company, Pearl River, New York

INTRODUCTION

Liquid chromatography/electrochemistry (LCEC) is the fastest growing electroanalytical technique today. The combination provides a powerful approach for the specific determination of trace quantities of electrochemically active molecules. Figure 1 illustrate the components of a typical LCEC system.

In a single component sample, quantitation by spectral or amperometric determination is simple. In a complex mixture, however, the optical spectra or the voltammetric curves of the compounds tend to be spread out and overlap with one another. The individual components in the mixture can rarely be determined based on the composite spectra or voltammogram. Because of the unavailability of ideal detectors with absolute specificity for the analytes, chromatography is introduced to enhance the selectivity of nonideal detectors and to time-resolve the individual substances in a complex matrix.

Principles of Electrochemical Detection

Electrochemical detection is based on controlled potential amperometry. A predetermined potential difference usually between +1.3 and −1.2 V, dependent on the redox behavior of the substance to be detected, is applied between the reference (Ag/AgCl) and the working electrode(s). Simply stated, the applied potential serves as the driving force for the electrochemical reaction. As the potential of the working electrode relative to the reference becomes more positive, the surface becomes a better oxidant (electron sink). Since the rate of material conversion by the electrochemical reaction (moles per second) is proportional to the instantaneous concentra-

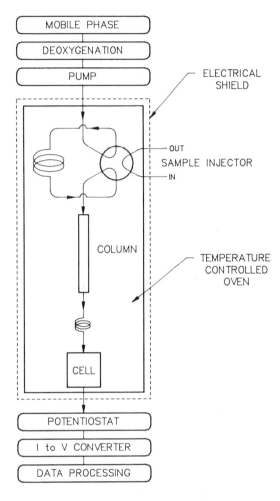

Figure 1 LCEC System. (Reproduced with permission of Bioanalytical Systems, Inc.)

tion, the current will be directly related to the amount of compound eluted as a function of time. If chromatographic conditions (mobile phase, flow rate, temperature, etc.) are carefully controlled, then amperometric detection is quite precise. Quantitative data can be obtained at the picomole level (total injected amount) for many compounds. Since only those molecules adjacent to the electrode surface are reacted, it is not surprising that the conversion efficiency (percent of reactant that actually reacts while passing over the electrode) is typically only 3–30% under normal operating conditions.

Liquid Chromatography/Electrochemistry

Frequently, the *amount reacted* is on the order of 10^{-15} mol. For example, in the case of a molecule with a molecular weight of 200 undergoing a two-electron transfer, only 5×10^{-13} g of sample might be converted into product for quantitation at a signal-to-noise (S/N) ratio of 5. Logically, it would seem worthwhile to increase electrode surface area and thereby increase the conversion efficiency. Unfortunately, the conversion efficiencies of both the analyte and the background electrolyte are increased, and the concomitant improvement in S/N ratio is not realized. In fact beyond a certain point, S/N decreases with increased area, for reasons outside the scope of this chapter. Working with low currents and small electrodes does require the ultimate low noise electronics with proper grounding and shielding.

The choice of the working electrode is critical to successful LCEC operation. Obviously, the surface should be physically and chemically inert to the mobile phase at the chosen applied potential. Two electrode surfaces that have found the greatest utility are carbon and mercury. The most versatile choice is glassy carbon. It has excellent resistance to nearly any solvent used in liquid chromatography and may be used over a wide potential range. Mercury, although it provides an extended negative range of potential, has very limited use in the positive direction. Conventional dropping mercury electrodes are not amenable to low dead volume cells. A better alternative is to employ a mercury film on a polished gold substrate. The mercury film can be made quite thin and very smooth. Mercury is better than glassy carbon when dealing with substances difficult to reduce, and it is the electrode of choice for many sulfur-containing compounds.

Electrochemistry in Thin Layers of Solution

The ideal flow cell for LCEC, which can faithfully reproduce the shape of concentration profiles eluting from high efficiency LC columns, is based on the thin-layer design [1]. Figure 2 is an illustration of such an amperometric transducer. The thin-layer channel is defined by a gasket held between a stainless steel block and a polymeric block. The stainless steel block is by itself an auxiliary electrode and is also the holder for the reference electrode. The polymeric block contains the working electrodes. The thin-layer cell can be easily disassembled for cleaning and replacement of electrodes.

To preserve the resolution of high efficiency microcolumns with chromatographic peak volumes well below 100 µL, detector cells can easily be made to have active volumes typically between 0.1 and 1 µL. The thin-layer electrochemical detector, with adjustable cell volume and tunable electronics, is versatile enough for all LC column formats in common use.

Further, the simultaneous use of two (or more) working electrodes can be facilitated to enhance both the qualitative and the quantitative aspects of an LCEC experiment [1, 2]. The detector can be changed from the "parallel mode" to the "series mode" of operation by rotating the working electrode block 90° relative to the flow stream. Regardless of the mode of operation, electrodes of the same or different materials, shapes, and surface areas may be used. The potentials on the individual electrodes may be independently controlled.

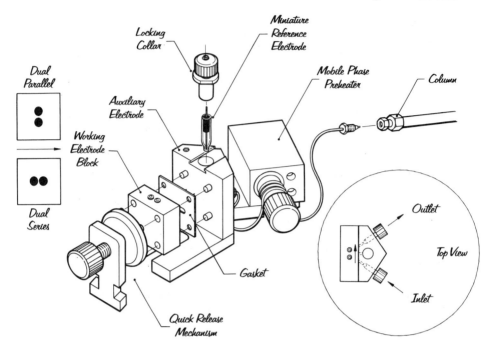

Figure 2 Thin-layer amperometric transducer. (Reproduced with permission of Bioanalytical Systems, Inc.)

For the parallel mode of operation, in which the analyte passes over all the electrodes simultaneously, the following advantages are realized:

1. Confirming peak identity and purity by the ratio of currents monitored at the individual electrodes
2. Enhancing the detection of the desired species by plotting the differences between two signals to subtract out "common mode" information
3. Improving selectivity for compounds with several redox states by applying an oxidizing potential to one working electrode and a reducing potential to the other electrode
4. Expanding the range of detection by simultaneously recording the signals from low and high potential reactions

A parallel-dual application, using response ratios at two potentials for confirming the identity of neurotransmitters and the related metabolites, is shown in Figure 3.

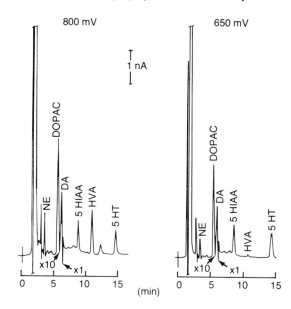

Figure 3 Determination of neurochemicals in rat brain homogenate using parallel dual electrode LCEC. (Reproduced with permission of Bioanalytical Systems, Inc.)

In the series mode of operation, products of reaction from an electrode upstream can be detected by a second electrode placed downstream, i.e., products of an oxidation reaction upstream are detected in the reductive mode downstream, and visa versa. Besides the enhanced selectivity as listed above (1 and 2) for the parallel mode, the following benefits accrue:

1. Enhancing the selectivity of the downstream electrode by discriminating against compounds that reacted irreversibly at the upstream electrode
2. Improving the selectivity and detectability of the compound of interest by monitoring at the downstream electrode a derivative formed at the upstream electrode
3. Discriminating against dissolved oxygen

A series dual application illustrating the enhancement of selectivity achieved at the downstream electrode is shown in Figure 4.

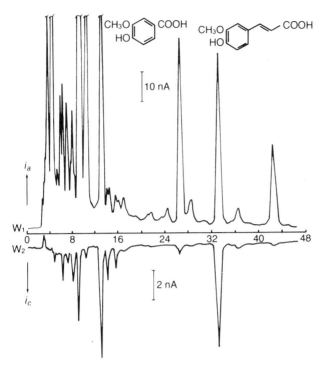

Figure 4 Series dual electrode chromatography of an extract of beer containing phenolic acids. The anodic chromatogram is plotted upwards and the corresponding cathodic chromatogram is plotted downwards. The cathodic chromatogram illustrates the enhancement of selectivity achieved at the downstream electrode. (Reproduced with permission of Bioanalytical Systems, Inc.)

ELECTROACTIVE DRUGS AND REACTION SCHEMES

Overview

As in the spectroscopic methods, molecular structure is the primary determinant for the electroactivity of an analyte. The accessibility of various filled and unfilled molecular orbitals ultimately determines the thermodynamics and kinetics of the electrode process. As for spectroscopy, there is a great body of empirical information that can be drawn upon to predict the behavior of individual compounds. In examining a candidate drug or metabolite, several key questions must be answered. What functional groups are present? Does the parent structure permit delocalization of the added positive or negative charge? Are there substituents present in the molecule that enhance or detract from electroactivity? Is the redox reaction pH dependent? What is the solubility? All of these factors are important considerations in assessing electroactivity for LCEC or any other electrochemical technique. Among the many electroactive organic compounds, the following classes of substances are frequently ideal candidates for LCEC: phenols (especially hydroquinones and catechols), aromatic amines, thiols, nitrocompounds, and quinones. There are many other classes that fit into these classifications, as well as some unique compounds such as ascorbic acid, uric acid, phenothiazines, and NADH. Organometallics such as *cis*-platinum anticancer agents are also amenable to electrochemical detection, along with other inorganic species, such as iodide and bisulfite, which may be present in bulk or formulated drug substances.

Precolumn and postcolumn derivatizations are now being utilized to make some compounds better candidates for LCEC. Chemical, electrochemical, enzymatic, and photolytic derivatizations in conjunction with both oxidative and reductive detection schemes are being used to determine species with normally unfavorable redox properties.

Cyclic voltammetry rapidly provides useful information preliminary to LCEC [3]. This convenient stationary-solution experiment easily duplicates those conditions (electrode material, electrolyte, etc.) found in the LCEC detector cell. Analogous to generating a UV spectrum, cyclic voltammetry elucidates oxidation and/or reduction behavior as a function of applied potential.

Figure 5 illustrates a cyclic voltammogram for the easily oxidized drug α-methyldopa. The cyclic voltammogram may be considered the electrochemical equivalent of an optical spectrum and is used in a similar manner. As the UV absorbance spectrum is used to choose an optimum wavelength for detection, the cyclic voltammogram is used to help choose an optimum potential for detection. The axes on which voltammograms are plotted can be confusing to the novice. Figure 6 illustrates the usual arrangement. Note that the energy axis should be viewed on a relative rather than an absolute basis, since the zero is established arbitrarily by choice of the reference electrode used [1].

Selectivity in LCEC

Using liquid chromatography in conjunction with an amperometric detector offers the advantages of efficient separation on microparticulate packing materials and the low detection limits (sub-picomole range) and selectivity of electrochemical techniques. While the current LC column technology is at a point where one may quickly achieve excellent separation of closely related substances, this is more often than not insufficient. This is especially true when a determination involves a biological matrix, such as blood or urine. Often selectivity must be enhanced by a cleanup step prior to injecting material on the column. This typically consists of one or more liquid-liquid or liquid-solid extractions carried out in a batch mode.

The selectivity of the column is also commonly augmented by the detector. This is a real strength of electrochemical detection schemes. While a "universal" detector could be useful in solving some simple problems, a "selective detector" is much more applicable in difficult determinations. The electrochemical detector is a tunable device that permits enhancement of selectivity by changing the applied poten-

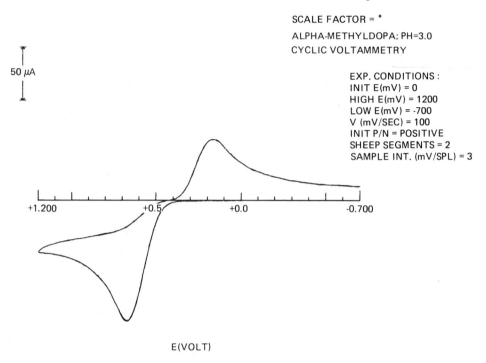

Figure 5 Cyclic voltammogram for α-methyldopa. (Reproduced with permission of Bioanalytical Systems, Inc.)

tial. The choice of electrode material, derivatizing agent, electrode configuration, and mobile-phase composition can also influence selectivity. To fully understand how this works, it is important to be familiar with hydrodynamic voltammetry, a technique that is also the basis for amperometric and coulometric titrations [1]. Hydrodynamic voltammograms are plotted on the same axes as are cyclic voltammograms. The difference results from the fact that the solution is moving, and the products of the electrode reaction are rapidly swept away from the electrode and therefore are not readily determined by reversing the potential scan. Hydrodynamic voltammograms (HDV) are generally developed by making repeated injections of a standard solution and stepping the potential of the detector between these injections. The chromatographically assisted HDV may be used in tandem with cyclic voltammetry to determine the optimum detector operating potential. In addition to matching retention times, these current-potential curves (HDVs) can be compared. In this way, identities of peaks are confirmed based on the chromatographic and electrochemical properties of a compound.

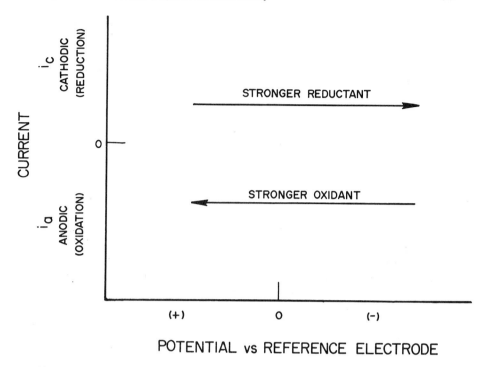

Figure 6 Potential controls position of equilibrium between O and R. (Reproduced with permission of Bioanalytical Systems, Inc.)

Representative Selection of Electroactive Drugs

Oxidative Applications (4)

Phenolic substances. These are for the most part readily oxidized at a graphite electrode. The oxidation potentials for phenols vary widely with structure, and some (hydroquinones and catechols) are far more readily oxidized than others. Many compounds of pharmacological interest (catecholamines, pharmaceuticals) and industrial interest (antioxidants, antimicrobials, agricultural chemicals) are phenolic, and trace determinations based on LCEC are now quite popular. Aldomet (α-methyldopa) is a hydroquinone (a catechol) representative of the more easily oxidized phenols. It undergoes a clean two-electron oxidation to an *o*-quinone:

Aromatic amines. Like phenols, aromatic amines are oxidized at a graphite electrode over a wide range of oxidation potential. Some compounds (phenylenediamines, benzidines, and aminophenols) are ideal candidates due to their very low oxidation potentials. The analgesic acetaminophen is an ideal example of an aminophenol that readily oxidizes to an acetylated quinone imine:

$$\text{acetaminophen} \longrightarrow \text{acetylated quinone imine} + 2e + 2H^+$$

Thiols ("sulfhydryls" or "mercaptans"). These compounds are very easily oxidized to disulfides in solution, but this thermodynamically very favorable redox reaction occurs only very slowly at most electrode surfaces (e.g., glassy carbon). Therefore, LCEC methods for thiols usually depend on the unique behavior of these compounds at a mercury electrode surface at about +0.10 V (a very low potential). The reaction involves formation of a stable complex between the thiol and mercury. In a formal sense, the mercury and not the thiol is oxidized:

$$2\ RSH + Hg \longrightarrow Hg(RS)_2 + 2e + 2H^+$$

Thiols are very susceptible to oxidation by dissolved oxygen. It is therefore not surprising that samples and mobile phases must be deoxygenated to achieve good precision and the lowest detection limits. This requires a chromatograph free of teflon tubing.

This approach has been used to determine, among others, the amino acid cysteine, the tripeptide glutathione, and the pharmaceuticals penicillamine and captopril. Besides thiols, many other sulfur-containing compounds are good candidates for LCEC.

Miscellaneous oxidizable compounds. A number of unique substances have been studied by oxidative LCEC. Being an excellent reducing agent, ascorbic acid is easily detected with excellent selectivity in very complex biological samples:

$$\text{ascorbic acid} \rightleftharpoons \text{dehydroascorbic acid} + 2H^+ + 2e$$

$$\text{dehydroascorbic acid} \xrightarrow{k} \text{diketogulonic acid} + H_2O$$

Liquid Chromatography/Electrochemistry

As for thiols, low detection limits require the use of deoxygenated mobile phase. Similarly, uric acid is readily detected in biological materials. The important enzyme cofactor NADH is readily oxidized at carbon electrodes and provides a unique opportunity for enzyme immunoassays coupled to LCEC:

$$\text{NADH} \longrightarrow \text{NAD}^+ + 2e + H^+$$

Some heterocycles of pharmacologic interest (phenothiazines, imipramine) are also uniquely applicable. Phenothiazines undergo a very clean one-electron oxidation to a cation radical:

$$\text{phenothiazine} \longrightarrow \text{phenothiazine}^{+\cdot} + e$$

Reductive Applications (4)

Quinones. These are among the best behaved organic compounds to undergo redox reactions in aqueous solutions. There are a reasonably large number of synthetic and natural products containing the quinone moiety, and many of these are excellent candidates for selective determination by LCEC. Unfortunately, some of the most important of these compounds (vitamin K_1) are extremely hydrophobic, due to the presence of long hydrocarbon side chains, and they are therefore quite difficult to study by reversed phase LC. A number of pharmacologic agents used in chemotherapy are quinones. For example, pharmacokinetic studies of the antibiotic doxorubicin can benefit from the following electrochemical reaction for use in an LCEC method.

$$\text{doxorubicin} + 2e + 2H^+ \longrightarrow \text{reduced doxorubicin}$$

Nitro and nitroso aromatic compounds. These have been among the most extensively investigated by both organic and analytical electrochemists. Aromatic nitro and nitroso compounds are very readily reduced at both carbon and mercury electrodes, but other compounds such as nitrate esters, nitramine, nitrosamines, and nitrosoureas are often good candidates as well. Often the selectivity is extremely good in biological and environmental samples because the nitro group is rare in nature, and few other organic compounds are so easily reduced. A good example is the popular antibiotic chloramphenicol, which can easily be determined in blood using a four-electron reduction to the corresponding hydroxylamine:

$$\underset{\substack{HCOH \\ | \\ HCNHC\overset{\diagup O}{\diagdown CHCl_2} \\ | \\ HCOH \\ H}}{\underset{NO_2}{\bigcirc}} + 4e + 4H^+ \longrightarrow \underset{\substack{HCOH \\ | \\ HCNHC\overset{\diagup O}{\diagdown CHCl_2} \\ | \\ HCOH \\ H}}{\underset{HNOH}{\bigcirc}} + H_2O$$

Reagents containing the aromatic nitro group have frequently been used to derivatize amines, aldehydes, and carboxylic acids, etc. to improve their characteristics for determination by absorption spectroscopy. The same or closely related reagents are now being used to provide an electrochemically active functionality for these types of compounds. Although it is true that aldehydes and ketones can be electrochemically reduced and alkyl amines and carboxylic acids can be electrochemically oxidized, the energy required to initiate these well-known reactions is at present far too great to permit development of a successful LCEC trace determination procedure without derivatization. This subject will be covered in some detail in the next section.

Miscellaneous reducible compounds. A number of organometallic compounds show promise for LCEC studies, and a few have already been examined in detail. More highly conjugated organic compounds, such as α, β-unsaturated ketones and imines, are occasionally good candidates, but at this time, ultraviolet (UV) detectors frequently outperform electrochemical detectors for such systems. To date, there have been relatively few reports on LCEC studies of metal ions in bulk pharmaceuticals or pharmacologically active metal complexes, compared to wholly organic species. The disulfides derived from in-vivo oxidation of thiol-based drugs can be detected by electrochemical reaction of the disulfide to the thiol and detection of the latter by oxidation as previously described. This approach uses the dual-series arrangement with two mercury film electrodes (see Figure 7).

Reaction Detection Schemes For LCEC

Introduction

In many cases, as in the determination of metabolites of neurochemical interest (e.g., tyrosine and tryptophan), analytes are electrochemically active and can be di-

Liquid Chromatography/Electrochemistry

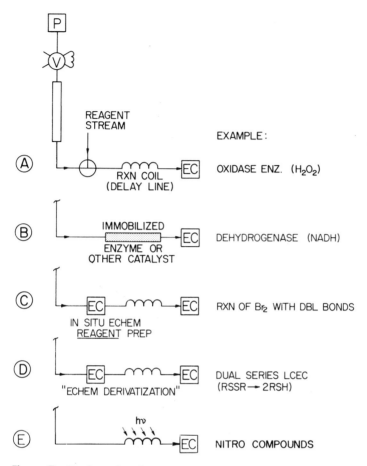

Figure 7 Configurations for postcolumn reactions with electrochemical analyzers. Various combinations of these five arrangements have also been used.

rectly detected at an electrochemical detector. Without a doubt, this fact is the primary reason the technique has become so popular in the last few years, in contrast to other methodologies that require derivatization to extend the sensitivity or increase the selectivity of a detection scheme following separation.

Simple assays requiring as few steps as possible are always the most reliable. Nevertheless, there are some problems in which derivatization appears to be the only practical possibility. Amino acids, with few exceptions, are universally determined by derivatization whether the technique be optical absorbance, fluorescence, gas chromatography, liquid chromatography, or thin-layer chromatography. Fatty acids (e.g., protaglandins), alcohols, alkylamines, and many thiols also present a challenge for most analytical methods. This challenge is often met by various forms of derivatization (chemical, photochemical, enzymatic, and electrochemical) followed by electrochemical detection.

Precolumn Derivatization For LCEC

Precolumn derivatization techniques are rapidly expanding in utility and popularity for improved detection following liquid chromatography. Most often these approaches have involved off-line, bulk chemical reactions, since these involve a lesser amount of time up front devoted to development of on-line reactor vessels (coils, columns, etc.) and optimization of the kinetics of these reactions in a flowing stream.

A wide variety of reagents have been used in the development of a precolumn derivatization techniques for LCEC (see Table 1 for a representative selection). Nitroaromatic compounds, such as 2, 4-dinitrofluorobenzene and trinitrobenzene sulfonic acid, used as derivatizing agents because of their excellent optical properties, have also found utility in LCEC applications. The nitro group is easily reduced to a hydroxylamine:

$$\phi NO_2 + 4e + 4H^+ \rightarrow \phi NHOH + H_2O$$

Due to the interference of oxygen in the use of reductive detection, a series-dual detector is often used to determine the hydroxylamine at very low positive potentials following its conversion to the nitroso derivative:

$$\phi NHOH \rightarrow \phi NO + 2e + 2H^+$$

Because the nitro group is quite unusual in biology, it provides a very selective handle for LCEC of many nonelectroactive compounds. Derivatives with more than one nitro group add sensitivity (more electrons transferred per molecule). Several good examples have appeared in the literature, including determination of GABA in biological media [5, 6], ketosteroids [7], amino acids [8], and peptides.

Due to the aforementioned interference of oxygen in the determination of compounds that have been reacted with a nitroaromatic compound, reagents that allow detection of substrates by oxidative LCEC are rapidly becoming the most widely utilized precolumn derivatizing agents. Currently, the most popular of these is o-phthalaldehyde (OPA). In the presence of a suitable thiol, primary amines react to form the 1-(S-alkyl)-2-alkylindole derivative.

The isoindole product is readily oxidized electrochemically, providing a very useful approach to a variety of alkylamines and sympathomimetic drugs [9], amino acids [10–12], and small peptides [11]. The application to biomolecules is of special sig-

Liquid Chromatography/Electrochemistry

nificance, since the OPA derivatives synthesized with t-butylthiol are more stable than the mercaptoethanol derivatives used in fluorescence applications [10, 11] and may yield an approximate twofold improvement in electrochemical response. Where OPA derivatization has shown limited utility in the determination of peptides by LC with fluorescence detection [13] due to poor quantum yield, electrochemical response exhibits a lesser structural dependence [11]. Results appear encouraging for determination of peptides and amino acids in biological samples using OPA and several other reagents (see Table 1).

Postcolumn Derivatization For LCEC

Postcolumn chemical reactions coupled to electrode reactions are also expanding the range of LCEC applications. Figure 7 illustrates five configurations. In 7a, a reagent is added, mixed, and reacted in a delay line, followed by electrochemical detection. A superb example of this is the determination of acetylcholine in brain tissue by LCEC [21]. Acetylcholinesterase and choline oxidase are mixed in the reaction coil, and the detection process proceeds as follows:

$$\text{Acetylcholine} \rightarrow \text{Choline} + \text{acetic acid}$$
$$\text{Choline} \rightarrow \text{Betaine} + H_2O + 2H_2O_2$$

The peroxide is detected at a platinum electrode. There are many pairs of esterases and oxidases that have been used to develop classical assays. Now many of these are being directly coupled to LCEC. Determination of reducing carbohydrates by similar postcolumn reaction schemes have been accomplished using copper (II)-1, 10-phenanthroline chelate [22] and 2-cyanoacetamide [23]. Dopamine-O-sulfate conjugates were determined as free dopamine after hydrolysis in a reaction coil [24]. The dopamine is then determined at the EC detector providing a much lower detection limit for metabolites. Such applications show the increasing utility of EC techniques in the study of metabolism.

In 7b, a catalyst is immobilized and a cofactor or other reaction product is detected. For example, the determination of acetylcholine described above may be accomplished in such a manner by coimmobilization of the esterase and oxidase [25]. In a similar manner, using a dehydrogenase enzyme, an alcohol can be detected indirectly by monitoring the turnover of NAD to NADH, the latter being ideal for electrochemical detection [26]. Dalgaard and coworkers have demonstrated the determination of both phenolic [27] and cyanogenic [28] glycosides using enzymatic postcolumn cleavage followed by EC detection. For the cyanogenic glycoside determination, a combination of the schemes in 7b and 7a is used. The cleavage occurs in the enzyme column:

$$R_1R_2C(CN)(O - Gly) \xrightarrow{\text{glycosidase}} R_1R_2C(CN)OH + GlyOH$$

which is followed by a base-catalyzed cleavage in a reaction coil:

Table 1 Selection of Reagents for Precolumn Derivatization

Reagent	Analyte	EC detection mode	Reference
Trinitrobenzene sulfonic acid	Amines, amino acids, γ-aminobutyric acid	Reductive	5, 6, 8
2,4-Dinitrofluorobenzene	Amines, amino acids	Reductive	8, 14
2-Chloro-3,5-dinitropyridine	Amines, amino acids	Reductive	8, 14
3,6-Dinitrophthalic anhydride	Peptides	Reductive	15
2-Carboxy-4,6-dinitro fluorobenzene	Peptides	Reductive	16
p-Nitrophenylhydrazine	Ketosteroids	Reductive	7
o-Phthalaldehyde	Amines, amino acids	Oxidative	9, 10, 11, 12
N-(4-anilinophenyl) malemide	Thiols (cysteine), glutathione, D-penicillamine	Oxidative	17
p-Aminophenol	Fatty acids, bile acids, prostaglandins	Oxidative	18
Phenylisothiocyanate	Amino acids	Oxidative	19
Enzyme multiplied imunoassay, NAD+/NADH	Phenyltoin	Oxidative	20

$$R_1R_2C(CN)(OH) + OH^- \rightarrow R_1R_2CO + CN^- + GlyOH$$

and finally detection of the CN^- at a silver electrode:

$$Ag + 2CN^- \rightarrow Ag(CN)_2^- + e^-$$

The automated method provided the first selective detection of cyanogenic glycosides.

In 7c, an upstream electrode generates a reagent (e.g., Br_2 from Br^- in the mobile phase, Ag^+ from a silver electrode) that reacts with a nonelectroactive compound (e.g., an unsaturated fatty acid), and the *decrease* in reagent concentration is monitored downstream [29]. In 7d, the analytes of interest are converted at an upstream electrode into a product that is more selectively detected downstream. Examples already mentioned include reduction of a nitro compound to a hydroxylamine and reduction of a disulfide to a thiol. Krull and coworkers have developed a modification of this technique by replacing the first electrode with a UV lamp in a quartz tube and performing photochemical/photolytic derivatization as illustrated in 7e [30]. This photolysis unit allows the direct formation of electroactive species from what are often nonelectroactive analytes. The technique has been demonstrated for the trace analysis of drugs and biologically active materials such as β-lactam antibiotics (penicillins, cephalosporins), barbiturates, benzodiazepines, and cocaine [30].

APPLICATIONS OF LIQUID CHROMATOGRAPHY/ ELECTROCHEMISTRY TO PHARMACEUTICAL RESEARCH

Overview

A phenomenal number of individual analytes of interest to researchers in all aspects of the pharmaceutical industry have been determined using liquid chromatography/ electrochemistry. In a single paper by Massart et al. [31], the minimal detectable concentrations (electrochemical versus absorbance detection) for 72 different drugs were compared. The classes of pharmacologic compounds included local anesthetics, antipyretics, tricyclic antidepressants, sulfonamides, sex hormones, beta-adrenoceptor blocking agents, phenothiazines, alkaloids, diuretics, and penicillins. Electrochemistry is the fastest growing detection technique for liquid chromatography. Its greatest merit has become obvious—the enhanced *selectivity*, and the improved detection limits that it affords, allow for a simplification of sample cleanup in biological applications. The ramifications of this are additional time and cost savings to the laboratory. Since a greater awareness of the utility of LCEC has evolved, a greater emphasis is being placed on LCEC techniques and applications in the development of pharmacologically active agents. Several excellent supplemental reviews are available that treat in more detail LCEC applications for individual drug classes and substances [4, 32].

Bulk Drug Analysis

The initial development of a pharmaceutical agent usually involves the development and characterization of the bulk agent. Although this is not an area in which LCEC is typically applied, recent applications have begun to demonstrate its utility, especially in the determination of the purity of reference standards.

Liquid chromatography followed by controlled potential coulometry has been utilized to determine the purity of electroactive drugs, such as acetaminophen [33]. A coulometric detector made of crushed reticulated vitreous carbon was used, and the number of coulombs generated were counted. By relating the number of coulombs generated in the electrochemical reaction to that which is predicted by theory, a method for the assignment of purity has been established. In this method, the faraday, 96,487 C/equiv., becomes a universal standard. The utility of the electrochemical detector as a screen for electroactive impurities has also been demonstrated [34]. Even when investigating the purity of a bulk drug that is not itself electroactive, the presence of electroactive impurities can lend important information on causes of contamination or routes of degradation. The combination of UV and electrochemical detection is becoming routine. Often UV is used to quantitate major components, while EC selectively detects impurities (or metabolites).

An important new area of pharmaceutical research is the development of biotechnology products. The lack of previous background information or therapeutic applications have led to stringent restrictions on trace levels of impurities in these bulk drug substances. Therapeutic applications of recombinant DNA products has been questioned for the trace levels of nucleic acids derived from recombinant bacterial chromosome and plasmid. Previous techniques for such low level determinations have been time consuming, laborious, and prone to experimental error. Kafil et al. have developed a highly sensitive and precise method for quantitation of DNA/RNA fragments [35, 36]. The method is based on hydrolysis of nucleic acids followed by LCEC. Previous investigations have indicated the utility of electrochemical detection for the quantitation of guanine nucleotides [37]. Using a dual electrode cell with a metal oxide electrode as the catalyst generator and a glassy carbon electrode as the detector, this method allows for quantitation of nucleic acids to the 50 pg level and DNA to the 100 pg level. Figure 8 illustrates the determination of DNA near the detection limit. The method allows for the determination of RNA in the presence of DNA in complex recombinant DNA and provides qualitative information on the adenine/guanine ratio. With the rapid expansion of biotechnology, it appears that the selectivity and sensitivity of LCEC will be applicable to bulk substances as well as determinations in biological samples.

A heretofore relatively little used application of LCEC is determination of the smaller anionic species such as cyanide and metabisulfite in bulk drugs. Simultaneous determination of cyanide, sulfide, iodine, and bromide by ion chromatography followed by electrochemical detection at a silver electrode has been demonstrated [38, 39]. Such rapid and sensitive determinations are especially applicable to bulk drug and formulated drug preparations where metabisulfite is sometimes added as an oxidant and the other species are considered contaminants.

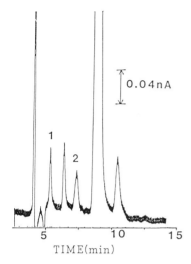

Figure 8 Chromatogram representing the determination of DNA near the detection limit: (1) guanine, (2) adenine. The original solution contained 1.0 ng/mL of *E. coli* DNA before hydrolysis. Injection volume: 100 L. (From Ref. 36, reproduced with permission of Bioanalytical Systems, Inc.)

Determination of Drugs in Biological Matrices

Analgesics

As a group, analgesics have been the drugs most widely studied by LCEC techniques. They are typically oxidizable aromatic compounds, either phenols and/or amines. The most widely studied of this group have been acetaminophen [40, 41, 46], salicylates [42], and morphine [43-46]. This also illustrates the most widely used application of LCEC for research purposes: determination of drugs in biological samples, such as blood, plasma, urine, organ tissues, and subcellular fractions (e.g., microsomes). This has led to further applications, such as the study of the metabolism of biologically active compounds, which will be discussed in more detail under "LCEC Approaches to the Study of Drug Metabolism."

Figure 9 illustrates the determination of a variety of analgesics in human serum. Demonstration of electroactivity of some of the more widely used and accepted analgesics has led to the use of the technique in the determination of the more recently discovered analgesics. Ciramadol and dezocine are synthetic opioid analgesics of the agonist-antagonist type and have phenolic moieties, which allow selective detection in plasma at the 10 μg/mL level [45]. In a similar manner, the narcotic analgesic, ketobemidene (1-[4-(3-hydroxyphenyl)-1-methyl-4-piperidyl]-1-propanone) is another new narcotic analgesic and phenolic that is amenable to electrochemical detection at the 1 ng/mL range in plasma [46]. LCEC has also been investigated in more detail recently as a method for opiate alkaloids of abuse including heroin, co-

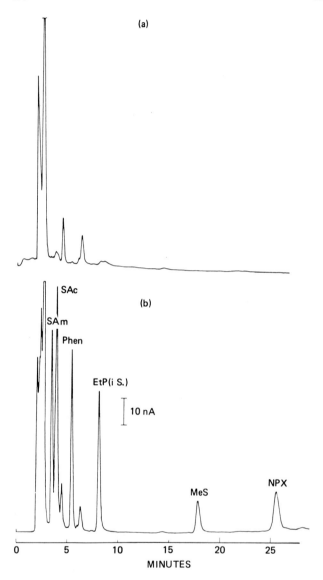

Figure 9 Isocratic LC separation of several analgesics in a spiked serum sample: (a) serum blank; (b) serum sample. Mobile phase = 0.05 M ammonium monochloracetate (pH = 3.2), 0.1 M NaClO$_4$ in MeOH:H$_2$O (50:50), Biophase ODS column. Detector potential = +1.15 V. Approximately 40 ng of each injected: salicylamide (SAm), salicyclic acid (SAc), phenacetin (Phen), ethylparaben (EtP), methylsalicylate (MeS), naproxen (NPX). EtP is used as an internal standard. (Reproduced with permission of Bioanalytical Systems, Inc.)

caine and related compounds [47, 48]. In the development of a general method [48], it is the oxidation of the aliphatic tertiary nitrogen atom common to these compounds rather than the phenolic moiety that is reacted at a glassy carbon electrode at +1.2 V versus Ag/AgCl. Detection limits were reported to be in the nanogram range for morphine, heroin, and cocaine.

Antibiotics

Traditionally, antibiotics have been analyzed by some form of bioassay that has specificity and that suffers interferences from both active metabolites and other antibiotics. Methods employing liquid chromatography with UV detection can suffer from poor detection limits and extensive sample preparation. While many antibiotics are *not* electrochemically active, several elegant applications have been developed.

Chloramphenicol is an antibiotic that can be assayed conveniently by LCEC at negative potentials at mercury film electrodes, like other compounds containing an aromatic nitro group. Since there are very few reducible compounds found naturally in blood, this assay is highly selective [49]. Furthermore, this compound may be determined using a series dual potential detection, with the upstream electrode set at −0.85 V (reducing −NO$_2$ group to −NHOH) and the downstream electrode set at +0.50 V (oxidizing −NHOH to −NO). This would allow determination without need for mobile phase deoxygenation.

Both trimethoprim and sulfonamides can be detected utilizing LCEC methodology at a greater sensitivity than is possible with UV detection, although the applied potential is high, generally about +1.1 to +1.2 V. Tetracyclines have several aromatic moieties which make them much ore amenable to electrochemical detection (typically about +0.60 V) [50]. Other antibiotics determined have included erythromycin and related macrolide antibiotics [51,52], enviroxime [53], several anticancer drugs (see the following section on chemotherapeutic agents), and amoxicillin [54].

While applications of electrochemical detection of antibiotics are not yet extensive, there are a number of applications including photolytic derivatization of β-lactams [28].

Chemotherapeutic Agents

Many of the diverse chemical agents used to treat various cancers have favorable redox properties for very selective LCEC methods. *Cis*-platinum complexes have been widely studied by LCEC, with various degrees of success. These species have been detected at several types of electrodes following LC including dropping mercury and hanging mercury drop electrodes using differential pulse amperometry [55–57], thin layer Au/Hg electrodes [58, 59], glassy carbon [58, 59] electrodes, and platinum electrodes utilizing a halide-catalyzed oxidation [60] (Figure 10). Although detection limits tend to be very dependent on ligands and also higher (typically in the micromolar range) than usually expected with EC detection, most of these species lack significant UV activity, and LCEC provides an excellent handle.

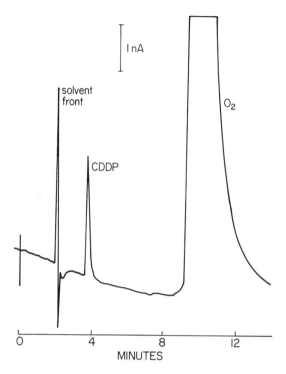

Figure 10 *cis*-Platin (CDDP) standard. A typical reductive LCEC chromatogram at the 0.63 ppm concentration level, using a reverse-phase C_{18} analytical column, with a mobile phase of 0.01 M acetate buffer (pH = 4.6) plus 0.15 mM HTAB. The Au/Hg working electrode is operated at an applied potential of -0.1 V. Under these LCEC conditions, there is an observed retention time of 3.6 min, with oxygen well resolved and appearing at about 9.33 min. The minimum detection limit for *cis*-Pt, using a signal to noise ratio of 3:1, has been determined to be about 10 ppb. (From Ref. 56, reproduced with permission of Bioanalytical Systems, Inc.)

The anthracyclines including daunorubicin and adriamycin [61-63] and related quinoid anti-cancer agents such as bisantrene, mitoxantrone [61, 62], and aziridinylquinones [64] are all quite easily determined in biological fluids using LCEC. The quinone moieties on these compounds are all easily reduced to the hydroquinones. Other common anticancer drugs such as methotrexate [65], procarbazine hydrochloride [66], and teniposide (a semisynthetic plant derivative) [61,67] are also easily determined with LCEC.

Thiols and Disulfides

As described previously, there are some functional groups that are uniquely suited to the LCEC approach. Thiols (mercaptans, sulfhydryls) and the corresponding disulfides form some of the most important redox couples in biochemistry (e.g., glutathione and glutathione disulfide). There are relatively few drugs containing these groups; however, captopril and penicillamine have received considerable attention in recent years. Such compounds can be very difficult to deal with due to their instability in the presence of dissolved oxygen. This problem is more often responsible for establishing a methods detection limit than is the instrumentation itself. With proper sample handling, LCEC would appear to be the method of choice for these compounds [68–70].

In most cases, the drugs have been detected directly; however, the instability problem can be partially overcome by derivatization [17]. N-(4-dimethylaminophenyl) maleimide was coupled to the thiol (captopril) using the traditional 1,4 addition reaction, protecting the thiol from oxidation. The tertiary amine on the derivatizing agent is electrochemically oxidizable at a more positive potential than the thiol. In this approach, one sacrifices the selectivity inherent in the low thiol oxidation potential in order to avoid decomposition of the analyte. The development of analytical methods often involves such compromises. Direct detection of the thiol usually proves to be satisfactory. Mercury electrodes generally give the best detection limits and selectivity, but glassy carbon can also be used [71]. Figure 11 depicts the simultaneous determination of captopril and its corresponding disulfide using a series dual detector with mercury film electrodes. This technique is easily applicable to other sulfhydryl drugs.

Additional Applications

A large number of diverse pharmaceuticals are nitrogen heterocycles, and they are unique both from pharmacological and from electrochemical points of view. The calcium antagonist nifedipine is a

Nifedipine

good example. The heterocyclic ring can undergo a two-electron oxidation (as per NADH), and the other ring bearing the nitro group can be easily reduced. Nifedipine and its metabolite have been determined with LCEC using both approaches [72–79].

Physostigmine is another unique structure that acts as a rather strong cholinesterase inhibitor. A series dual electrode approach to its determination in plasma [75] is based on the oxidation of

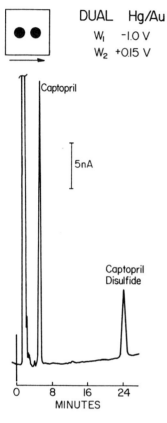

Figure 11 Captopril in urine. Dual Hg/Au chromatogram of acidified urine spiked with captopril (10 µg/mL) and captopril disulfide (20 µg/mL). The urine was prepared by adding 200 L of 1M HCLO containing 3 g/L Na₂EDTA to 1 mL of urine. The very clean appearance of the complex urine matrix is due to the high degree of selectivity inherent in the dual Hg/Au detector for thiols and disulfides. After being chromatographically resolved, these compounds pass into the dual detector where the disulfides are reduced at the upstream electrode, and the thiols are simply detected at the downstream electrode. The result is a single chromatogram depicting both thiols and disulfides. (Reproduced with permission of Bioanalytical Systems, Inc.)

Physostigmine

physostigmine at +1.0 V. The oxidized form rapidly hydrolyzes to form a quinoneimine product that can be reduced (2 electrons) at +0.1 V [76] (Figure 12). This scheme allows for a detection limit of 0.5 ng/mL for a 2 mL plasma sample.

Many other classes of compounds are detectable with electrochemical methods. They include beta-mimetics and beta-blockers (e.g., ritodrine, terbutaline, nylidrin, and metaproternol), L-dopa and other biogenic amines (e.g., levodopa and carbidopa), tricyclic antidepressants (mianserin, imipramine, and desipramine), phenothiazines (prochlorperazine, trimeprazine, neuroleptics, and antihistamines) [32], antiasthmas (e.g., theophylline, piriprost potassium) [77, 78], and cardiotonics (enoximone, piroximone) [79, 80]. This document touches on only a few of the staggering number of pharmacologically active agents that are amenable to LCEC.

LCEC Approaches to the Study of Drug Metabolism

Overview

The metabolism of drugs and other xenobiotic compounds (those species that by definition serve no nutritional or biochemical role in vivo) is frequently ideal to study from an electrochemical perspective because the most interesting pathways (from a toxicological viewpoint) often lead to products with lower redox potentials than the initial reactant. This means that minor drug metabolites can often be detected by LCEC at lower concentrations than the drug itself. Due to the complex nature of the samples required to follow such biological transformations and the extremely low levels at which both drug and metabolites are often present, this makes LCEC and excellent approach to a challenging analytical problem. The development of an LCEC method for such a task will ultimately allow for rapid trace determinations of analyte at varying levels of system complexity. The ability to determine the rate of reaction of a drug and the chemical nature and yields of specific products can also provide information on the mechanism of metabolism. These procedures can then be applied to the subsequent reactions of activated metabolites with target nucleophiles that may lead to toxicity.

LCEC is particularly suited to study the transformation of a drug and the subsequent metabolism of primary metabolite(s). Dual electrode amperometric detectors have led to several advantages. The parallel-adjacent mode may be utilized to confirm peak identity based on the electrochemical characteristics of a compound [81,

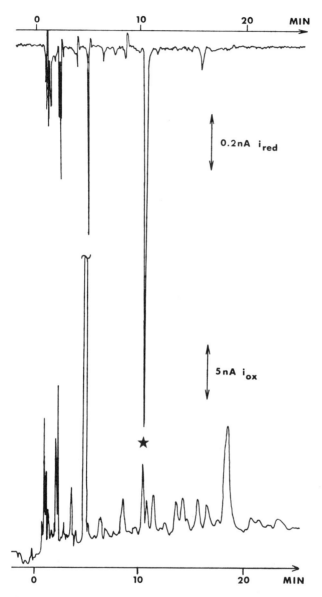

Figure 12 Physostigmine in blood plasma. Dual series chromatogram of plasma spiked with physostigmine (14 ng/mL). The plasma is prepared by addition of an ion-pairing reagent, followed by cleanup on a C_{18} Sep-Pak™ by elution with water and 60% methanol and final collection in 1 mL of methanol. After chromatography, the physostigmine is oxidized (i_{ox}) at +1.0 V vs. Ag/AgCl to allow formation of the quinoneimine which is reduced (i_{red}) at the downstream electrode. (Reproduced with permission of Bioanalytical Systems, Inc.)

82] and also to monitor to redox states simultaneously [83]. The series configuration has been used to determine important endogenous thiols [4] and to identify metabolites based on their electrochemical reversibility [84]. This section will describe applications of LCEC in the important areas of Phase I and Phase II drug metabolism.

Determination of Endogenous Electroactive Compounds

Biogenic amines. The influence of a drug or a disease state on the metabolic functions of an organism is an important aspect in the determination of drug efficacy. The first widespread use of LCEC was the determination of biogenic amines and a host of other important phenolics in various biological media [4].

Currently, it is advances in chromatographic technology that are improving the manner in which these compounds are determined using EC detection. "Microbore" techniques are distinguished by the low dead volume needed in the chromatographic system (especially the detection system) and the low flow rates used. Thin-layer electrochemical cells with low dead volumes (less than 0.3 µL) are ideal [84, 85]. The efficiency of the surface reaction in amperometric detection increases with decreasing flow rates, and since microbore columns allow for elution of much more concentrated segments detection limits for some biogenic amines have been reported to be below 1 femtomole [86]. With cartridge column techniques (columns ranging anywhere from 1 to 5 cm in length) and using modified thin-layer cells with fast electronic current amplifiers, a series of eighteen biogenic amines and metabolites may be separated in 4-7 min from brain tissue [87].

Future developments in this arena will probably lie in sampling technology. As in the study of all metabolic transformations, whether they are endogenous or exogenous analytes, the ideal technique would be to monitor changes in vivo using a totally noninvasive technique. Some strides are being made in this area using in vivo voltammetry, but since compounds may exhibit similar electrochemical properties, interference among analytes is a strong possibility. The technique of microdialysis is a new methodology that allows for the monitoring of chemical events in extracellular space, where chemical transmission takes place [88- 91]. It is an extension of the push-pull technique, except that the perfusing fluid is circulating inside a semipermeable membrane instead of freely in the tissue (Figure 13), in effect acting as a synthetic "blood vessel." Fluid is pumped through the probe at a slow rate (1-10 µL/min), and dialysate from the extracellular media is transported directly to the LCEC system. Since the system is used with awake, unrestrained animals, a true profile of metabolic changes over time may be realized. A number of physiologically important substances may be determined in the perfusate, and the comparison of in vitro techniques with in vivo experiments is a real possibility. Since a great number of exogenous substances such as drugs also enter into the extracellular fluid, microdialysis should also find application in xenobiotic metabolism studies. By uniting microdialysis with LCEC, the number of analytes that may be determined far exceeds that attainable using other noninvasive techniques such as biosensors.

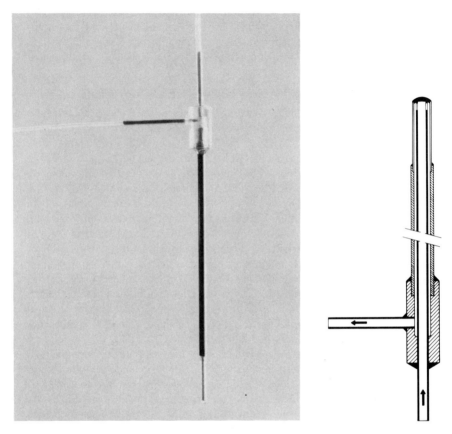

Figure 13 (Left) Photograph of a microdialysis probe as described in the text. The diameter of the stainless steel shaft is 0.64 mm and the diameter of the membrane is 0.5 mm. (Right) Schematic of the microdialysis probe showing its principal features. The dimensions are different from the photograph on the left in order to make the construction more obvious. Liquid enters through a steel cannula at the top leaving it at the bottom of the probe streaming upwards in the space between the inner tube and outer membrane. This is where the dialysis takes place. The liquid leaves the probe by way of the lateral cannula. (Reproduced with permission of Bioanalytical Systems, Inc.)

Liquid Chromatography/Electrochemistry

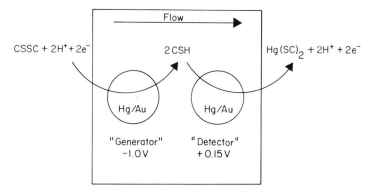

Figure 14 Schematic of twin Hg/Au series electrode arrangement for conversion of cystine to cysteine on line and its subsequent downstream measurement as cysteine. (From Ref. 90, reproduced with permission of Bioanalytical Systems, Inc.)

Thiols and disulfides. Liquid chromatography/electrochemistry exhibits a distinct advantage over chemical or enzymatic methods of analysis for thiols and disulfides, in that the two components may be determined simultaneously rather than by difference technique, and the method is applicable to all compounds with these functional groups. The series dual electrode technique utilizes mercury amalgamated gold electrodes (Figure 14). At the upstream or "generator" electrode disulfides are reduced to thiols at a negative potential of -1.0 V; all thiols (endogenous and reduced disulfide) are then detected at the downstream or "detector" electrode via the reaction with mercury [92]. This technique has been used to monitor various thiols and disulfides including glutathione, cysteine, homocysteine and their disulfides in urine [92, 93], and tissue samples and homogenates [94, 95] (Figure 15). Such compounds are indicated in the secondary metabolism of many drugs and xenobiotics, kidney disorders, and such genetic disorders as cystinuria and homocysteinuria.

Additional applications. Other endogenous analytes have been determined using LCEC techniques including prostaglandins [29], uric acid, ascorbic acid, NADH [4], tocopherols, vitamin K, ubiquinols [97], and estradiols [98, 99]. Pterins, nitrogen-containing heterocyclics, may be determined simultaneously in many biofluids using parallel adjacent dual electrodes even though as a class they exist in various oxidation states [83]. Peak confirmation was made for these species based on ratio of response at the electrochemical detector. Such current rationing techniques have also been used for identification and determination of quanine nucleotides [37]. This varied group of materials indicates the vast number of determinations and assays to which LCEC may be applied for endogenous materials.

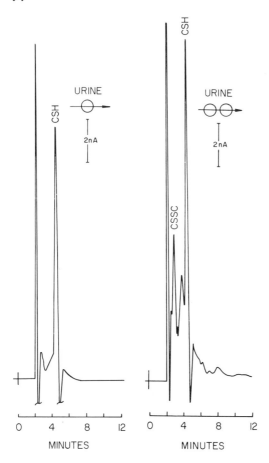

Figure 15 (a) Series dual Hg/Au electrode detection of the thiol cysteine in urine with only the downstream electrode on. (b) Simultaneous determination of cysteine and the disulfide cystine in urine using the dual series configuration with the upstream electrode now on. Column: Biophase C-18, 5 μ, Mobile phase: 96% monochloroacetic acid buffer (pH = 3), with 1 mM sodium octyl sulfate; 4% methanol. Flow: 1 mL/min. Injection 20 μL. Mobile phase is deoxygenated by N_2 purging. (From Ref. 96, reproduced with permission of Bioanalytical Systems, Inc.)

Determination of Primary and Secondary Drug Metabolites

Drug metabolizing enzymes generally catalyze the biotransformation of most often lipophilic aryl compounds to more polar, water-soluble products that are more readily excreted. The metabolism of these tissue-penetrating, lipophilic species is categorized by two phases [100]. Phase I reactions (which include the mixed function oxidations such as cytochrome P-450) catabolize the molecules by oxidation, reduc-

Liquid Chromatography/Electrochemistry

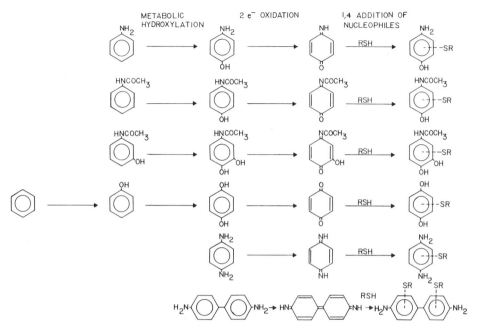

Figure 16 Reactions of some aryl xenobiotics, including acetaminophen (4-hydroxyacetanilide), benzene, and phenol.

tion, and hydrolysis. These reactions introduce polar reactive groups to the molecule, groups that increase water solubility and create a species more reactive towards the second metabolic phase. In Phase II, the xenobiotic or its Phase I products combine with endogenous substrates (glucuronate, sulfate, acetate, glycine, glutathione, etc) making them more readily excretable [101]. In general, the Phase I reactions that make many aryl based aromatic drugs more readily excretable also make them electroactive.

Evidence has been building that indicates that certain aryl xenobiotics may undergo direct two-electron oxidations in vivo (Figure 16). Such conversion may be mediated by the mixed-function oxidase and other enzyme systems. This electron withdrawal supersedes the normal oxygen insertion mechanisms and creates electrophilic species. A case in point is the formation of the quinoid N-acetyl-p-quinoneimine from acetaminophen [102]. This compound and similar species may undergo rapid 1,4 Michael type additions with endogenous nucleophiles. Some enzyme systems, such as the peroxidase, catalase, and prostaglandin H synthase, oxidize via one-electron withdrawals, the way an analyte is reacted at an electrode surface. With such mimicry between electrochemical and metabolic reactions, it is not surprising that LCEC is readily applicable to the determination of primary (Phase I) and secondary (Phase II) metabolites and may also lend insight to the occurrence of transformations.

Liquid chromatography/electrochemistry has been applied to the determination of metabolites from several sources of varying levels of system complexity. Using pure enzyme incubations and subcellular fraction incubations (microsomes), those substances needed for conjugation are eliminated. However, the matrix allows for easy determination of initial metabolic products. Higher levels include tissue slices, perfusates, urine, and plasma, for all of which LCEC has shown utility because of the selectivity and sensitivity advantages [102-111]. These advantages notwithstanding, dual electrode detection has allowed scientists to perform electrochemical experiments on injected solutions to characterize metabolites [102-104, 109-111]. Since even small changes in chemical structure will influence the potential of a compound, oxidation/reduction parallel adjacent detection mode may be used to confirm peak identity and more importantly allow an evaluation of peak homogeneity in these complex samples [109-110]. The series detector has, as previously described, been most often used as a postcolumn reactor, followed by an electrochemical detector. For the identification and quantitation of metabolites, tentative structural as-

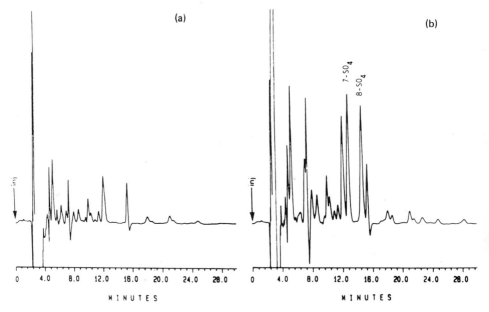

Figure 17 Determination of 7-and 8-sulfate conjugates in human urine after ingestion of fenoldopam: (a) chromatogram of urine prior to ingestion of fenoldopam. (b) 0-24 hr after ingestion of 100 mg fenoldopam. The response is measured at the second working electrode of a series configuration detecting the quinones formed at the first electrode. (From Ref. 104, reproduced with permission of Elsevier Science Publishers.)

signments may be made based on the reversibility of the electrochemical reaction. In the determination of fenoldopam (a renal vasodialator) and its metabolites, series detection was used to determine catechol sulfate and methoxy conjugates [104]. The conjugates oxidized at a potential of +0.9 V, which did not allow sufficient sensitivity for determination. However, at this potential, the conjugate is cleaved to form an o-quinone that is determined by reduction at the second electrode at a potential of 0.0 V (Figure 17). This technique should have general utility for drugs or xenobiotics in which one phenolic group is conjugated.

The ability to perform experiments as described above on the chromatographic time scale makes LCEC an invaluable resource for the scientist involved in drug metabolism studies.

REFERENCES

1. P. T. Kissinger, ed. *Introduction to Detectors for Liquid Chromatography*, BAS Press, West Lafayette, Ind. (1981).
2. C. E. Lunte, R. E. Shoup, P. T. Kissinger, *Anal. Chem.*, 57: 1541 (1985).
3. P. T. Kissinger and W. R. Heineman, eds. *Laboratory Techniques in Electroanalytical Chemistry*, Marcel Dekker, New York (1984).
4. R. E. Shoup, *Recent Reports on Liquid Chromatography with Electrochemical Detection*, Bioanalytical Systems, West Lafayette, Ind. (1982).
5. W. L. Caudill, L. A. Papach, and R. M. Wightman, *Current Separations*, 4: 59 (1982).
6. W. L. Caudill, G. P. Houck and R. M. Wightman, *J. Chromatogr.*, 227: 331 (1982).
7. K. Shimada, M. Tanaka, T. Nambara, *Anal. Lett.*, 13: 1129 (1980).
8. W. A. Jacobs and P. T. Kissinger, *J. Liq. Chromatogr.*, 5: 889 (1982).
9. P. Leroy, A. Nicolas, and A. Moreau, *J. Chromatogr.*, 282: 561 (1983).
10. L. A. Allison, G. S. Mayer, and R. E. Shoup, *Anal. Chem.*, 56: 1089 (1984).
11. W. Jacobs, *Current Separations*, 7: 39 (1986).
12. M. H. Joseph and P. Davies, *J. Chromatogr.*, 227: 125 (1983).
13. T. M. Joys and H. Kim, *Anal. Biochem.*, 94: 371 (1979).
14. W. A. Jacobs and P. T. Kissinger, *J. Liq. Chromatogr.*, 6: 125 (1983).
15. K. H. Xie, S. Colgan, and I. S. Krull, *J. Liq. Chromatogr.*, 6: 125 (1983).
16. J. L. Meek, *J. Chromatogr.*, 226: 401 (1983).
17. K. Shimada, M. Tanaka, and T. Nambara, *Anal. Chim. Acta.*, 147: 375 (1983).
18. S. Ikenoya, O. Hiroshima, M. Ohmae, and K. Kawabe, *Chem. Pharm. Bull.*, 28: 2941 (1980).
19. R. R. Granberg, *LC Magazine*, 2:776 (1984).
20. H. M. Eggers, H. B. Halsall, and W. R. Heineman, *Clin. Chem.*, 28:1848 (1982).
21. P. E. Potter, J. L. Meek, and N. H. Neff, *J. Neurochem.*, 41: 188 (1983).
22. N. Watanabe and M. Inoue, *Anal. Chem.*, 55: 1016 (1983).
23. S. Honda, T. Konishi, and S. Suzuki, *J. Chromatogr.*, 299:245 (1984).
24. M. A. Elchisak, *J. Chromatogr.*, 255: 475 (1983).
25. G. S. Mayer, *Current Separations*, 7: 47 (1986).
26. S. Kamada, M. Maeda, A. Tsuji, Y. Umezawa, and T. Kurahashi, *J. Chromatogr.*, 239: 773 (1982).
27. L. Dalgaard, L. Nordholm, and L. Brimer, *J. Chromatogr.*, 265: 183 (1983).

28. L. Dalgaard and L. Brimer, *J. Chromatogr.*, *303*: 67 (1984).
29. W. P. King and P. T. Kissinger, *Clin. Chem.*, 26: 1484 (1980).
30. I. S. Krull, C. M. Selavka, R. J. Nelson, K. Bratin, and I. Lurie, 1986 International Electroanalytical Symposium, Abstract No. 24 (1986).
31. G. Musch, M. DeSmet, and D. L. Massart, *J. Chromatogr.*, *348*: 97 (1985).
32. C. Lavrich and P. T. Kissinger, Liquid Chromatography-Electrochemistry: Potential Utility for Therapeutic Drug Monitoring (S. H. Y. Wong, ed.), Marcel Dekker, New York, p. 191 (1985).
33. G. W. Schieffer, *Anal. Chem.*, *57*: 968 (1985).
34. D. M. Radzik, 1986 Pittsburgh Conference on Analytical Chemistry and Applied Spectroscopy, Abstract No. 963 (1986).
35. J. B. Kafil, H. Y. Cheng, and T. A. Last, *Anal. Chem.*, *58*: 285 (1986).
36. J. B. Kafil and T. A. Last, 1986 International Electroanalytical Symposium, Abstract No. 33 (1986).
37. T. Yamamoto, H. Shimizu, T. Kato, and T. Nagatsu, *Anal. Biochem.*, *142*: 395 (1984).
38. R. D. Rocklin and E. L., Johnson, *Anal. Chem.*, *55*:4 (1983).
39. A. M. Bond, I. D. Heritage, G. G. Wallace, and M. J., McCormick, *Anal. Chem.*, *54*: 582 (1982).
40. D. J. Miner and P. T. Kissinger, *J. Pharm. Sci.*, *68*(1): 96 (1979).
41. J. W. Munson, R. Weierstall, and H. B. Kostenbauder, J. Chromatogr., *145*: 328 (1978).
42. D. A. Meinsma, D.M. Radzik, and P. T. Kissinger, *J. Liq. Chromatogr.*, *6*: 2311 (1983).
43. J. E. Wallace, S. C. Harris, and M. W. Peek, *Anal. Chem.*, *52*: 1328 (1980).
44. J. A. Owen and D. S. Sitar, *J. Chromatogr.*, *276*: 202 (1983).
45. A. Locniskar and D. J. Greenblatt, *J. Chromatogr.*, *374*: 215 (1986).
46. P. A., Hynning, P. Anderson, U. Bondesson, and L. Boreus, *J. Chromatogr.*, *375*: 207 (1986).
47. T. D. Wilson, *J. Chromatogr.*, *301*: 39 (1984).
48. R. S. Schwartz and K. O. David, *Anal. Chem.*, *57*: 1362 (1985).
49. Bioanalytical Systems, West Lafayette, Ind., LCEC Application Note No. 36, Chloramphenicol.
50. M. A. Alawi and H. A. Ruessel, *Chromatographia*, *14*(12): 704 (1981).
51. G. S. Duthu, *J. Liq. Chromatogr.*, *7*: 1023 (1984).
52. G. S. Duthu, 1985 International Electroanalytical Symposium, Abstract No. 37 (1985).
53. R. J. Bopp and D. J. Miner, *J. Pharm. Sci.*, *71*(12): 1402 (1982).
54. M. A. Brooks, M. R. Hackman, and D. J. Mazzo, *J. Chromatogr.*, *210*: 531 (1981).
55. S. J. Bannister, L. A. Sternson, and A. J. Repta, *J. Chromatogr.*, *273*: 301 (1983).
56. W. J. F. van der Vijgh, H. B. J. vender Lee, G. J. Postma, and H. M. Pinedo, *Chromatographia*, *17*: 333 (1983).
57. F. Elferink, W. J. F. van der Vijgh, and H. M. Pinedo, *J. Chromatogr.*, *320*: 379 (1985).
58. I. S. Krull, X. D. Ding, C. Selavka, and F. Hochberg, *Current Separations*, *4*: 14 (1982).
59. I. S. Krull, X. D. Ding, S. Braverman, C. Selavka, F. Hochberg and L. A. Sternson, *J. Chromatogr. Sci.*, *21*: 166 (1983).
60. W. N. Richmond and R. P. Baldwin, *Anal. Chim. Acta*, *154*: 133 (1983).

61. J. A. Sinkule, C. Akpofure, and W. E. Evans, *Current Separations,* 4: 68 (1982).
62. C. Akpofure, C. A. Riley, J. A. Sinkule, and W. E. Evans, *J. Chromatogr.,* 232: 377 (1982).
63. J. Pursley, *Current Separations,* 4: 1 (1982).
64. R. Driebergen, W. vanOort, P. Zuman, S. Postma, W. Verboom, and D. Reinhoudt, 1985 International Electroanalytical Symposium, Abstract No. 23 (1985).
65. J. Lankelma and H. Poppe, *J. Chromatogr.,* 149: 587 (1978).
66. R. J. Rucki, A. Ross, and S. A. Moros, *J. Chromatogr.,* 190: 359 (1980).
67. P. Canal, C. Michel, R. Bugat, G. Soula, and M. Carton, *J. Chromatogr.,* 375: 451 (1986).
68. Bioanalytical Systems, West Lafayette, Ind., LCEC Applications Note No. 48, Measuring Penicillamine in Plasma and Urine.
69. Bioanalytical Systems, West Lafayette, Ind., LCEC Applications Note No. 47, Captoril in Plasma.
70. L. M. Shaw, H. S. Bonner, and J. Nakamura, 1986 International Electroanalytical Symposium, Abstract No. 7 (1986).
71. I. C. Shaw, A. E. M., McClean, and G. H. Boult, *J. Chromatogr.,* 275: 206 (1983).
72. K. Bratin and P. T. Kissinger, Current Separations, 4: 4 (1982).
73. H. Suzuki, S. Fujiwara, S. Kondo, and I. Suggimoto, *J. Chromatogr.,* 341: 341 (1985).
74. N. D. Heubert, M. Speeding, K. D. Haegele, *J. Chromatogr.,* 353: 175 (1986).
75. K. Isakksson and P. T. Kissinger, *J. Chromatogr.,* 419:165 (1987).
76. X. Chen, P. He, and P. T. Kissinger, *J. Electroanol. Chem.,* 284: 371 (1990).
77. M. S. Breenberg and W. J. Mayer, *J. Chromatogr.,* 169: 321 (1979).
78. R. C. Lewis and F. J. Schwende, 1985 International Electroanalytical Symposium, Abstract No. 34 (1985).
79. A. G. Hayes, S. Mehta, and T. Chang, *J. Chromatogr.,* 336: 446 (1984).
80. C. L. Housmyer and R. L. Lewis, 1986 International Electroanalytical Symposium, Abstract No. 47 (1986).
81. S. M. Lunte and P. T. Kissinger, *Chem. Biol. Interaction.,* 47: 195 (1983).
82. G. S. Mayer and R. E. Shoup, *J. Chromatogr.,* 255: 533 (1983).
83. C. E. Lunte and P. T. Kissinger, *Anal. Chem.,* 55: 1458 (1983).
84. M. Goto, T. Nakamura, and D. Ishii, *J. Chromatogr.,* 226: 33 (1981).
85. E. J. Caliguri and I. N. Mefford, *Brain Res.,* 296: 156 (1984).
86. T. Huang, R. E. Shoup, and P. T. Kissinger, *Current Separations,* 9:139 (1990).
87. P. Y. T. Lin, M. C. Bulawa, P. Wong, L. Lin, J. Scott, and C. L. Blank, *J. Liq. Chromatogr.,* 7: 509 (1984).
88. C. B. Kissinger and P. T. Kissinger, *American Laboratory,* March 1990, p. 94.
89. U. Ungersted, Chapter 4, *Measurement of Neurotransmitter Release In Vivo,* (C. A. Marsden, ed.), John Wiley, New York, p. 81 (1984).
90. C. C. Louillis, J. N. Hingtgen, P. A. Shea, and M. H. Aprison, *Pharmacol. Biochem. Behav.,* 12: 959 (1980).
91. J. B. Justice, Jr., S. A. Wages, A. C. Michael, R. D. Blakely, and D. B. Neill, *J. Liq. Chromatogr.,* 6: 1873 (1983).
92. L. A. Allison and R. E. Shoup, *Anal. Chem.,* 55: 8 (1983).
93. S. M. Lunte, *Current Separations,* 6: 30 (1984).
94. D. C. Sampson, P. M. Stewart, and J. W. Hammond, *Biomed. Chromatogr.* 1:21 (1986).

95. E. G. DeMaster, F. N. Shirota, B. Redfern, D. J. W. Goon, and H. T. Nagasawa, *J. Chromatogr.*, *308*: 83 (1984).
96. S. M. Lunte and P. T. Kissinger, 1984 International Electroanalytical Symposium, Abstract No. 36 (1984).
97. J. Lang, K. Gohil, and L. Packer, Tenth International Symposium on Column Liquid Chromatography, San Francisco, Abstract No. 2405 (1986).
98. K. Shimada, F. Xie, and T. Nambara, *J. Chromatogr.*, *378*: 17 (1986).
99. C. Bunyagidj and J. A. McLachlan, 1986 International Electroanalytical Symposium, Abstract No. 43 (1986).
100. M. Ingelman-Sundberg, Bioactivation or Inactivation of Toxic Compounds, *Drug Metabolism and Distribution*, (J. W. Lamble, ed.), Elsevier Biomedical Press, Amsterdam, p. 22 (1983).
101. D. V. Parke, *Biochem. Soc. Trans.*, *11*: 457 (1983).
102. D. J. Miner and P. T. Kissinger, *Biochem. Pharmacol.*, 28: 3285 (1979).
103. D. A. Meinsma and P. T. Kissinger, *Current Separations*, 6: 42 (1985).
104. V. K. Boppana, F. C. Heineman, R. K. Lynn, W. C. Randolf, and J. A. Ziemniak, *J. Chromatogr.*, *317*: 463 (1984).
105. K. Shimosato, M. Tomita, and I. Ijiri, *J. Chromatogr.*, *377*: 279 (1986).
106. M. J. Cyronak, K. M. Dolce, V. K. Boppana, and J. A. Ziemniak, 1986 International Electroanalytical Symposium, Abstract No. 24 (1986).
107. K. McKenna, P. J. Kraske, and A. Brajter-Toth, 1986 International Electroanalytical Symposium, Abstract No. 24 (1986).
108. M. A. Brooks and G. DiDonato, *J. Chromatogr.*, *337*: 351 (1985).
109. D. A. Roston and P. T. Kissinger, *Anal. Chem.*, *54*: 429 (1982).
110. S. M. Lunte and P. T. Kissinger, *Chem. Biol. Interactions*, *47*: 195 (1983).
111. J. R. Rice and P. T. Kissinger, *Biochem. Biophys. Res. Comm.*, *104*: 1312 (1982).

5
Radiochemical Quantitation: Considerations for HPLC

Jeff Quint

Beckman Instruments, Inc., Fullerton, California

John F. Newton

SmithKline Beecham Pharmaceuticals, Swedeland, Pennsylvania

INTRODUCTION

Isotopically labeled drugs provide, through the use of a variety of radiochemical techniques, highly specific and sensitive mechanisms for quantitation. However, the presence of a radiolabel alone may not provide sufficient specificity, especially when used in a biological system. In many cases, radiolabeled metabolic products can interfere in the quantitation of the drug under study. A mechanism for resolving these problems is the use of an appropriate separation system in the quantitation of radiolabeled drugs. One of the most popular methods for separation and quantitation of radiolabeled drugs and putative metabolites involves the use of HPLC with radiochemical quantitation, by either discrete or flow-through methodology.

For many years there has been an interest in performing on-line radioactivity monitoring for chromatography systems [1,2]. This article will discuss some of the methods of detection used in flow counting for radio-HPLC, and its application to pharmaceutical research.

Discrete Counting (Fraction Collection)

Fraction collection with scintillation counting is still used in many laboratories for analysis of radiochromatography data. Fractions are collected at preset time intervals to divide the chromatogram into discrete sections. Depending upon the volume of the fractions collected, scintillation fluid can be added directly to the collection vial, or an aliquot can be taken from the collection vial and placed in an additional vial. Current liquid scintillation cocktails can accommodate a sample, especially of those which contain a reasonable percentage of organic, of equal volume without a significant drop in counting efficiency. Radiochemical content of samples is determined at time intervals that are dependent upon the required sensitivity of the assay. The radioactivity is plotted against fraction number to obtain a radiochromatogram, and peaks are integrated by lumping the fractions of interest together. This is a time-consuming process that often limits the usefulness of radiochromatography. Fortunately, many of the state-of-the-art LS counters available today have software that markedly reduces the time spent in data reduction. The resolution in this method is controlled by the size of the fractions collected. The sensitivity is controlled by the period of time that each fraction is counted.

Flow-Through Counting

Undoubtedly, the most popular technique of radiochemical detection of HPLC eluates involves the use of on-line flow-through radiochemical detectors. Depending upon the radionuclide, liquid or solid scintillators can be utilized. Most of the detection techniques for radioactivity are based on trapping the energy of the radioactive particle and converting that energy to light scintillations. The number of scintillations, measured as counts per minute (CPM), is proportional to the number of radioactive disintegrations (DPM). The ratio of CPM to DPM is the detector efficiency:

$$100 \times \frac{CPM}{DPM} = \text{Detector efficiency}$$

When using a discrete counter, the counting time is set by the investigator. With a flow detector, the counting time, known as the residence time, is the time that any radioactive molecule resides in the flow cell. The approximate residence time is dependent on the size of the flow cell and the flow rate as given by

$$\text{Residence time (min)} = \frac{\text{Flow cell volume(mL)}}{\text{Flow rate (mL/min)}}$$

Sensitivity in flow-through counting is dependent on flow rate and flow cell volume. When a scintillation cocktail is combined with HPLC effluent for the purpose of determining radiochemical content, the flow rates through the detector cell (liquid) are substantially greater than with a flow cell that is packed with solid scintillators; this dictates that liquid scintillation cells be 3–4 times larger than their counterparts containing solid scintillants in order to provide similar residence times. Residence times generally range from 0.05 to 1 min with flow-through radiochemical quantitation, which is a very short time in comparison to normal discrete counting. Residence time can be increased by slowing the flow rate or increasing the flow cell volume. However, it is not reasonable to expect detection and quantitation of peaks in a flow detector that normally requires counting times of 5-10 min per fraction in order to obtain adequate results.

Isotopes with High Penetrating Radiation

Isotopes whose radioactive particles are energetic enough to penetrate through tubing can be measured by an external detector without the aid of a scintillator within the flow cell. Devices commonly employed for detection of high penetrating radiation include Geiger–Müller counters or photomultiplier tubes (PMT) with crystal scintillators. These types of detectors (shown in Figure 1) are acceptable for such isotopes as ^{125}I, ^{131}I, ^{32}P, ^{36}Cl, and ^{99}Tc.

Isotopes with Low Penetrating Radiation

Isotopes with low penetrating power must be detected inside the tubing rather than externally. For example, tritium beta particles travel only a few microns in water. Therefore, in order to detect a radioactive disintegration, some of the energy of the beta particle must be trapped by the scintillator before it is lost to the solution. Low energy isotopes such as ^{3}H, ^{14}C, ^{35}S, and ^{45}Ca are usually detected by either a liquid scintillator or small particles of solid scintillator. Table 1 shows the penetrating power of various isotopes in liquid media and the approximate scintillation yield with liquid scintillation cocktail. The low scintillation yield observed for ^{3}H demonstrates the difficulties encountered in the detection of ^{3}H.

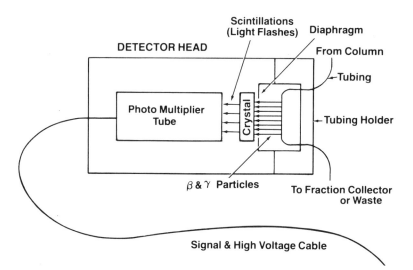

Figure 1 Detection of radioisotopes with high penetrating radiation. On-line detection is based upon the fact that the radioactive particles pass out of the effluent tubing into a crystal. This interaction causes the crystal to scintillate. These scintillations are recorded as radioactive counts.

Table 1 Penetrating Power of Beta Emitting Isotopes

Isotope	Energy (keV)	Distance of travel in water (microns)	Scintillation yield per disintegration (photons)
^{3}H	18	1	225
^{14}C	156	100	1,950
^{32}P	1,710	10,000	21,500

DETECTION METHODS

Liquid Scintillation

In liquid scintillation, an aromatic solvent is used to trap the energy of the beta particle and transfer this energy to a fluorescence molecule that produces a scintillation. Since most samples are in aqueous effluents, the aqueous liquid must be dissolved in a detergent that traps the sample in small micelles. The formation of the micelles is dependent on the composition of the HPLC effluent. High salt solutions and organic solvents, like methanol and acetonitrile, affect the proper formation of micelles and thus proper liquid scintillation counting.

The liquid scintillation process is affected by two types of interferences, chemical quench and color quench. Under conditions of chemical quench, the energy from the beta particle is trapped by the solution before a scintillation is produced. Under conditions of color quench, the scintillation occurs but the light is trapped by a colored complex. Changes in chemical quench, which are often dependent upon the HPLC effluent, can have a dramatic effect on the CPM observed by the detector (Table 2). Note that because of its low energy ^{3}H is dramatically affected by solvent composition, while the higher energy of ^{14}C is only slightly affected. However, the amount of chemical quench that results from a specific HPLC effluent is dependent upon the actual cocktail that is used (Table 3). Therefore, the selection of an appropriate cocktail can have a dramatic effect upon the counting efficiency obtained in flow-through radiochemical detection. Thus the efficiency obtained is dependent upon the HPLC effluent, the specific cocktail employed, and the exact ratio of each.

Under isocratic conditions, counting efficiency will remain relatively constant. The CPM results can be converted to DPM results by dividing by the constant counting efficiency. Under gradient conditions, the counting efficiency often varies with the solvent composition. This is corrected by running a constant amount of radioactivity in the cocktail through the system during a gradient analysis and storing a continuous function of counting efficiency versus retention time. The CPM may then be corrected to DPM by dividing by the counting efficiency at any given retention time.

The keys to efficient detection in liquid scintillation are proper selection of sample and cocktail compatibility, and proper mixing to insure a homogeneous solution. Sample and cocktail compatibility can be determined by mixing the two solutions at

Table 2 Effect of Chemical Quench on Counting Efficiency in Flow-Through Radiochemical Detection

Cocktail/effluent	^{3}H[a]		^{14}C[a]	
	Ready Flow III™	FloScint III™	Ready Flow III™	FloScint III™
Water	100	100	100	100
50% MeOH	92	90	98	98
80% MeOH	90	95	98	96
50% acetonitrile	75	75	99	89
80% acetonitrile	75	71	96	86
PO$_4$ buffer	98	98	98	100
TFA 1%	94	97	99	97
Ion pairing salts	88	95	99	96

[a]CPMs observed are expressed as a percentage of those observed when water alone was used as the HPLC effluent. The ratio of cocktail to effluent was maintained at 3:1.
Ready Flow III is a trademark of Beckman Instruments.
FloScint III is a trademark of Radiomatic Instruments.

Table 3 Tritium Efficiency as a Function of Cocktail Type in Chemical Quench[a]

HPLC Conditions	Cocktail[b]		
	Ready Solve CP™	FloScint III™	Ready Flow III™
80% acetonitrile	92	71	75
PO$_4$ buffer	100	98	98

[a]The tritium efficiency is expressed as a percentage of that observed using Ready Solve CP™ and phosphate buffer.
[b]The ratio of cocktail to HPLC effluent was maintained at 3:1.
Ready Solve CP and Ready Flow III are trademarks of Beckman Instruments.
FlowScint III is a trademark of Radiomatic Instruments.

different ratios to determine the point at which the resulting solution gels or becomes nonhomogeneous. Once a proper ratio of effluent and cocktail has been established under static conditions, it should be determined whether the ratio produces homogeneous solutions under flow-through conditions (i.e., whether mixing is complete). The production of homogeneous solutions under flow-through conditions is dependent upon the type of scintillant pump and the style of mixing tee used. If an even (nonpulsatile) flow is not achieved, proper mixing may not occur, which often results in cloudy or improperly mixed solutions. Furthermore, if the style of the mixing tee does not produce proper mixing, similar results are obtained. Changes in the style of mixing tee can render a nonhomogeneous solution homogeneous, which can increase tritium efficiency by 30%. Visual inspection of the cocktail/effluent mixture, after it passes out of the mixing tee, is undoubtedly the best documentation of mixture homogeneity in the flow-through situation.

Most flow cells for liquid scintillation are coils of Teflon tubing. The volume of the cell is regulated by the internal diameter and length of tubing. The volume of most cells is fixed, but in some flow cells the tubing length and diameter can be modified (Figure 2).

Figure 2 Liquid scintillation flow cell. This flow cell allows user installation of tubing for rapid change of flow cell size. The black key is used for winding tubing within the flow cell.

Radiochemical Quantitation

The effective limits of flow-through detection of radiochemicals are functions of both flow cell size and flow rate (cocktail and HPLC effluent) through the cell. Data presented in Table 4 exemplify the problems encountered in quantitation of low levels of radioactivity using liquid scintillation techniques. As the amount of radioactivity in the peak decreases, the variance observed in its quantitation increases significantly. This is especially apparent when the total radiochemical content of a peak is less than 1000 DPM.. The marked increase in variability in the quantitation of SKF 86002 shown in Table 4 is not observed when peak areas are quantitated by optical (UV) methodology. The large variability observed for low level quantitation of radioactivity can be reduced slightly by eliminating the variability introduced by the injector with the use of an internal standard (quantitation of the ratio of radioactivity to the internal standard).

The high variability observed at low levels of radioactivity is a function of the variability in counting statistics, which is inversely proportional to residence time. By decreasing the flow rate through the cell, it is possible to reduce the variability observed in quantitation of low levels of radioactivity, which lowers the effective limits of quantitation. In liquid scintillation techniques, the flow rate through the cell can be decreased primarily by a reduction in cocktail flow rate. However, a radical reduction (i.e., less than 2:1, cocktail:effluent) in cocktail flow rate, without a subsequent reduction in HPLC flow rate, results in a decrease in the cocktail:effluent ratio, which can adversely affect counting efficiency and, consequently, counting statistics. This effect could nullify the decreased variability in counting statistics gained by an increased residence time in the cell. Other techniques used to decrease the uncertainty in flow-through liquid scintillation include decreasing total flow rate through the flow cell (maintaining a constant effluent:cocktail ratio) and using a larger flow cell. Both techniques increase the residence time within the cell and decrease the variability in counting statistics.

Generally speaking, as the residence time within a cell is increased the resolution of the system is decreased. Therefore, decreasing the flow rate through the cell or increasing the cell size results in peak broadening and tailing. Techniques used to increase the resolution of a system generally include the opposite of those described above: (1) increasing the flow rate through the cell (in liquid scintillation techniques this is generally accomplished by increasing the cocktail:effluent ratio) and (2) decreasing the flow cell size. The effect of different flow rates (accomplished by altering the cocktail:effluent ratio) on the resolution of three lauric acid metabolites that differ only in the position of hydroxylation is displayed in Figure 3. A more dramatic effect on resolution is observed when the size of the flow cell is decreased (Figure 4).

In any flow-through liquid scintillation system, the resolution of a system must always be weighed against the limits of detection. If high resolution is required in a specific assay, it will be very difficult to attain high sensitivity also (with reasonable variability, CV <20%). As instrumentation becomes more sophisticated, the situation will improve, but at present the only means of meeting requirements for high

Table 4 Variability in the Quantitation of SKF 86002 by Radiochemical and Absorbance Techniques[a]

SKF 86002 conc. (µg/mL)	Radiochemical				Absorbance		
	Mean peak radioactivity DPM	Coefficient of variance (%)[b]		Mean peak area	Coefficient of variance (%)[b]	
		DPM	DPM/IS area[c]		Area	Area/IS area[c]
1.0	11843	2.91	3.02	103.97	0.52	0.65
0.5	6077	7.15	7.43	53.14	0.80	0.34
0.1	1100	9.14	8.84	9.98	2.87	2.58
0.05	663	26.15	21.60	4.88	2.40	2.39

[a]HPLC flow rate was 2.0 mL/min; cocktail flow rate was 6.0 mL/min. Flow cell volume was 400 µL.
[b]Coefficient of variation was determined by 5 replicate injections.
[c]Internal standard (SKF 85838) was quantitated by absorbance at 254 nm.

Radiochemical Quantitation

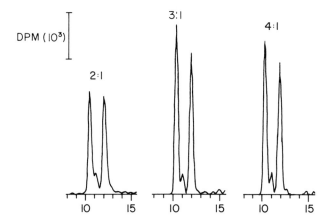

Figure 3 Separation of lauric acid metabolites by HPLC with radiochemical detection: effect of various scintillant flow rates. Metabolites were separated on an Altex 5 µM C_{18} (4.6 × 250 mm) column using a solvent system of 27% acetonitrile in 1% acetic acid. Scintillant (Ready Solve CP™) flow rate was increased from 3.0 mL/min (2:1 ratio) to 6.0 mL/min (4:1 ratio), while HPLC flow rate was maintained at 1.5 mL/min. A 1.0 mL flow cell was used for all analyses. (From Ref. 3, used with permission.)

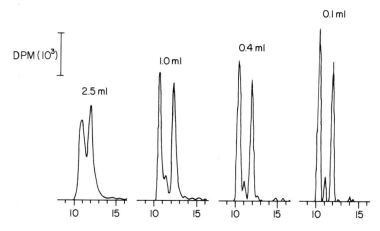

Figure 4 Separation of lauric acid metabolites by HPLC with radiochemical detection: effect of flow cell volume. Conditions are similar to those described in Figure 3. The scintillant (Ready Solve CP™) flow rate was maintained at 4.5 mL/min, and the HPLC flow rate was maintained at 1.5 mL/min, during all analyses. Actual flow cell sizes are indicated above. (From Ref. 3, used with permission.)

resolution and sensitivity in an assay is to increase the specific activity of the radiopharmaceutical under examination.

Solid Scintillation

The first solid scintillators contained large anthracene crystals. Since solid scintillators must trap the energy of the beta particle directly, these scintillators were only marginally useful for counting ^{14}C and of no use for ^{3}H. The detection efficiency of solid scintillators can be increased by decreasing the particle size, which reduces the spacing between particles and thereby increases the chances of a beta particle contacting the solid scintillator. The advantages of solid scintillators are that they are nondestructive to the sample, they are not affected by salts and solvent concentration changes, and they are very efficient for all common isotopes except tritium. The disadvantages are that samples may bind to the solid scintillator and that solid scintillators have a low counting efficiency for ^{3}H.

Most of the flow cells for solid scintillators are coiled Teflon tubing or U-shaped glass tubes containing the appropriate packing materials. These flow cells have fixed volume; a totally new cell must be purchased to change volumes. The introduction of straight tubes of Teflon or glass that can be removed and repacked by a dry packing procedure has significantly reduced the cost of using solid scintillators. Not only are the straight tubes reusable but the volume of the cell can be changed by changing the id of the tubing (Figure 5).

Table 5 shows some of the characteristics of solid scintillators. The most useful solid scintillators have proven to be yttrium silicate beads, which have excellent efficiency for ^{14}C and adequate efficiency for many ^{3}H applications. Because the energy of the beta particle is transferred directly to the solid scintillator, chemical quench is eliminated. Recent research into surface treatment of yttrium silicate has eliminated many of the common adherence problems. Research continues on surface treatments to minimize sample binding to the scintillator and on improvement of flow cell design for better ^{3}H efficiency.

Most of the considerations concerning resolution and efficiency that apply to liquid scintillation techniques also apply to solid scintillation techniques. However, flow cell size and design, such that laminar flow problems are minimized, are primary determinants of resolution and sensitivity when solid scintillation techniques are employed. Because scintillant is not added to the HPLC effluent in solid scintillation techniques, increasing or decreasing flow rate through the cell, in order to increase resolution or sensitivity, respectively, is not considered a viable option. In extreme cases, it is possible to add a carrier solvent at some position distal to the column but prior to entry into the radiochemical flow cell. This technique can be used to change sensitivity or resolution or to introduce a stronger solvent to prevent adsorption to solid scintillator particles.

Radiochemical Quantitation 111

Figure 5 Solid scintillator cell with replaceable scintillator cartridges.

Table 5 Characteristics of Solid Scintillators

Type	Size (microns)	^{3}H Efficiency (%)	^{14}C Efficiency (%)
Plastic beads	200	0.1	25
Glass (lithium activated)	70–80	0.1	25
CaF$_2$ (Eu)	100	3	90
Yt silicate	70–80	5	85

Selection of Solid Versus Liquid Scintillation

Hardware Considerations

It is important to optimize the hardware to match the scintillation event. During the liquid scintillation event, the duration of the fluorescence process is a few nanoseconds; in order to reduce the background of the system, a coincidence circuit is designed to accept only those pulses from the two phototubes that occur within 10–20 ns of each other. When a solid scintillator is used with a 10–20 ns coincidence gate, many of the real pulses are lost, because they occur up to 60 ns apart. This reduces the counting efficiency of the solid scintillator for ^3H and ^{14}C by approximately 50%. Modern flow detectors, therefore, adjust their circuits for differences between liquid and solid scintillations.

Scintillant Considerations

Since solid scintillators are easier to use, give similar efficiency when compared to liquid scintillators (except for ^3H), and are not destructive to the sample, most investigators prefer to use them. Since some compounds adhere to solid scintillators, one should evaluate whether a particular compound adsorbs to the scintillator under separation conditions. The following method may be used to check the compatibility of a sample with the scintillator:

1. Weigh 100 mg of scintillator into a 1.5 mL microcentrifuge tube.
2. Add 10–50 µL of the sample, enough for about 100,000 CPM.
3. Add 1000 µL of the solvent system being used during separation.
4. Cap and vortex the tube.
5. Spin in a microcentrifuge for 3–5 min.
6. Decant the supernatant into a scintillation vial.
7. Add counting fluid and count the eluent.
8. Repeat 3–7 with a stronger solvent to check for further recovery.
9. Repeat 2–7 to check for the total amount added to the tube.

The recovery of the radioactivity from the scintillator is determined by

$$\text{First recovery} = \frac{\text{DPM removed in wash [1]}}{\text{Total DPM added}} \times 100$$

If the first recovery is greater than 90%, then the sample is compatible with the scintillator. Table 6 shows results of this test with some radiochemicals and solvent systems. Greater than 90% recovery provides good results without the possibility of adsorption. If recovery is less than 90%, another type of solid scintillator may be tried or a liquid scintillator should be used. Alternatively, a stronger solvent may be introduced at a point distal to the column but prior to the radiochemical flow cell.

Table 6 Recovery from Yttrium Silicate Solid Scintillators

Compound	Eluting solvent	Recovery (%)
Lysine	0.1 M phosphate buffer	93.6
Glutamic acid	0.1 M phosphate buffer	94.0
Glucose	0.1 M phosphate buffer	89.3
UDP glucoronide	0.1 M phosphate buffer	91.7
Angiotensin I	0.1 M phosphate buffer	91.3
Cortisol	43% methanol 57% water	94.7
Dehydroepiandrosterone	59% methanol 41% water	94.0
Carbonic anhydrase	0.1 M phosphate buffer	2.0

Flow-Through Radiochemical Techniques: Preparative Techniques

Many new radiochemical detectors can control a fraction collector by changing fractions when a peak is detected. This is difficult to do correctly because the detector must discriminate between real peaks and noise. Discrimination is accomplished by continuous evaluation of the signal for two residence times after the sample passes through the detector in order to determine whether the signal is a peak or noise. There is a delay after detection of a peak before the fraction collector is changed to account for the length of tubing from the detector to the fraction collector.

By connecting the fraction collector to the detector, the detector can minimize the number of fractions collected and record how much radioactivity was put into each test tube. The amount of radioactivity per fraction can be documented in the printout at the termination of the run. An event mark on the analog printout of the radiochromatogram documents each fraction change. Some new detectors allow versatility in how peaks are fractionated and in the quantitation of each fraction by permitting selection of fraction times within peaks and outside of peaks.

Normally, the detection of radioactive compounds with liquid scintillation requires that the column effluent must be mixed with a cocktail during analysis. Therefore, in order to use liquid scintillation for preparative work, a portion of the HPLC effluent must be diverted to the fraction collector before it is mixed with the liquid scintillation cocktail. This reduces the limit of radiochemical quantitation. Under preparative conditions, the sensitivity of liquid scintillators will be similar to or lower than that of solid scintillators depending on the splitting ratio used. For example, if 20% of the sample is mixed with the liquid scintillation cocktail and 80% is directed to the fraction collector, the sensitivity of detection is five times lower.

Flow-Through Radiochemical Techniques: Quantitation

Data analysis for flow-through quantitation of radiochemicals is often quite sophisticated. The analog output for radioactive data is noisy because of the random nature of the radioactive signal (Figure 6). Simple smoothing routines can improve the signal, but the data is still not very useful. This type of smoothing also causes loss of resolution. By applying sophisticated digital filtering techniques, a reduction in noise can be achieved without loss of resolution (Figure 6).

Quantitation of the amount of radioactivity that elutes from an HPLC column can be calculated by most radiochemical detectors. By summing the number of observed counts from the start to the end of the peak, the radioactivity under the peak can be calculated. Some detectors keep track of peaks, retention time, and percent of total activity, as shown in Figure 7. Many of the modern detectors can plot data from several analog instruments. Therefore, output from two different detectors can be overlayed for comparison purposes. Such a system has great utility in complex matrices where it is important to correlate radiolabeled peaks with UV spectra. An example of the specificity gained in overlaying chromatograms is shown in Figure 8.

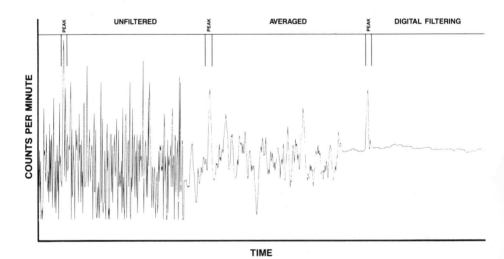

Figure 6 Effect of signal averaging and digital filtration on analog signals from a radiochemical detector. Because of the random nature of the radioactive decay process, analog signals continuously monitoring a flowing sample are very noisy (see unfiltered signal below). Averaging the signal helps to improve the analog output (see averaged signal). In many radiochemical detectors, the microprocessor uses a sophisticated digital filtering technique to improve the quality of the analog output.

FRACTION DATA

FRACTION NUMBER	PEAK NUMBER	ISOTOPE #1 14C (CPM)	STARTING FRACTION TIME (MINUTES)	FRACTION VOLUME (mL)	RCM (%)
1		**********	******	*****	
2		1035	0.018	0.57	0.09
3	1	17719	0.592	0.83	0.02
4		1698	1.419	0.70	0.07
5	2	27961	2.122	0.62	0.01
6	3	17393	2.742	0.74	0.02
7		583	3.486	0.21	0.06
8	4	8784	3.692	0.83	0.03
9		142	4.519	0.08	0.07
10	5	26858	4.602	1.07	0.02
11		1823	5.676	0.74	0.07
12	6	8213	6.420	0.87	0.03
13		119	7.288	0.04	0.08
14	7	15454	7.330	0.87	0.02
	TOTAL	127781		TOTAL 8.18	

PEAK DATA   14C

PEAK NUMBER	FRACTION NUMBER	ISOTOPE #1 (CPM)	RETENTION TIME (MINUTES)	% OF RUN
1	3- 3	17719	0.919	14.48
2	5- 5	27961	2.450	22.85
3	6- 6	17393	2.944	14.21
4	8- 8	8784	3.977	7.18
5	10- 10	26858	4.926	21.95
6	12- 12	8213	6.705	6.71
7	14- 14	15454	7.656	12.63
	PEAK TOTAL	122382		
	RUN TOTAL	127781		

Figure 7 Typical data summary from a flow-through radiochemical detector. These include the peak number, the activity in a peak, the retention time, and a percentage of total radioactivity that each peak accounted for.

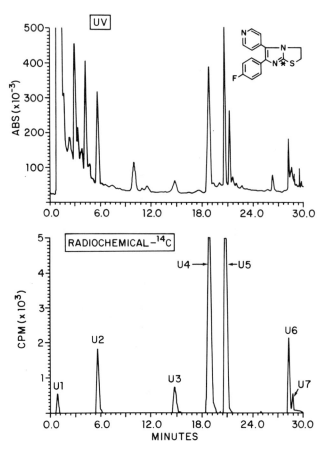

Figure 8 HPLC chromatogram of radioactivity and UV absorbance of urinary metabolites of SKF 86002. SKF 86002 and metabolites were quantitated on a 5 μM NOVAPAK C_{18} column (5 × 100 mm) using a gradient of 100 mM ammonium acetate and 75% acetonitrile in 50 mM ammonium acetate maintained at a flow rate of 1.8 mL/min. UV absorbance was monitored at 254 nm. The scintillant (FLO-SCINT II™) to HPLC effluent ratio was maintained at 3; flow cell volume was 400 μl.

Example of Radiochemical Quantitation in Pharmaceutical Analyses

Undoubtedly, metabolism studies constitute a large portion of radioisotope usage within the pharmaceutical industry. Radiolabeled drugs are often used for metabolism/pharmacokinetic studies in animals and man. In addition, radioisotopes are often used to quantitate the in vitro metabolism of pharmaceutical agents at the subcellular, cellular, or whole organ level. The primary reason for use of radioisotopes in drug metabolism studies is the ability to demonstrate, utilizing chromatographic techniques, and quantitate, sometimes unknown, metabolic products. In addition, when radiolabeled, both drugs and metabolites can be visualized and quantitated in extremely complex matrices.

While radiochemical quantitation provides many advantages for quantitation of drugs and their metabolites, there are several problems unique to radiochemical quantitation that must be taken into consideration. Probably the most important parameter to be documented in radiochemical quantitation of drugs and/or metabolites is the exact column recovery of injected radioactivity obtained. Normally, column recovery should range from 95-105%. When column recoveries are less than 95%, the primary reason, assuming no problems in injection, is adsorption of drug and/or metabolites to the column. This is observed most often with highly reactive compounds and/or metabolites. Column recoveries substantially greater than 105% are most commonly observed during long HPLC runs following injection of small amounts of radioactivity. Apparent recovery of more than 100% of injected radioactivity is the result of a small underestimation of background activity that is accentuated over a long HPLC run. If column recoveries for a mixture of drug and metabolites range between 95 and 105%, then it is legitimate to use column recovery to standardize actual injection volume.

Radiochemical quantitation following HPLC separation has been used in a variety of in vitro metabolism studies. It is not the intent of this review to detail each of these studies. Rather, a few will be referenced as examples of the versatility of radiochemical techniques in HPLC. One of the first demonstrated uses of HPLC with flow-through radiochemical detection was in the quantitation of the in vitro metabolism of progesterone and testosterone in testicular cells [4]. Flow-through technology has also been used to document in vitro enzyme activities. Many investigators have utilized lauric acid hydroxylation as a crude measure of the microsomal content of the cytochrome(s) P-452 [5,6]. HPLC coupled with flow-through radiochemical quantitation allows for quantitation of both (ω- and (ω-1)- hydroxylation reactions, which are supported by different isozymes of cytochrome P-450 [7]. Recent advances in our laboratory employing smaller flow cells and liquid/liquid flow-through detection have led to the demonstration of a third microsomal metabolite of laurate, (ω-2)-laurate [3]. A sample chromatogram of the in vitro metabolism of lauric acid as analyzed by HPLC with radiochemical detection is shown in Figure 9. The great resolving power of HPLC coupled with the specificity of radiochemical detection offers the opportunity to separate and quantitate metabolites that differ very little. An example of this is shown in Figure 10. This is a chromatogram of the products of the in vitro metabolism of SKF 102081, a leukotriene receptor antagonist [8]. These products differ only by the position of hydroxylation. This minor change, on a molecule of nearly 500 daltons, can easily be resolved by HPLC and quantitated using flow-through radiochemical detection.

In vivo studies utilizing radiolabeled pharmaceutical agents can be broadly grouped into three categories: (1) pharmacokinetic studies, (2) mass balance/excretion studies, and (3) tissues distribution studies. HPLC with radiochemical quantita-

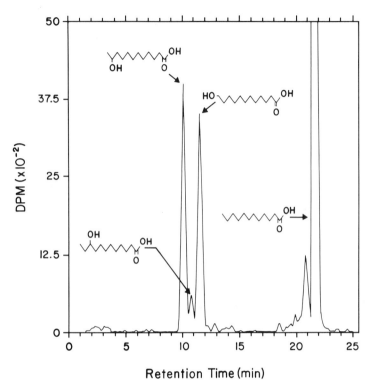

Figure 9 HPLC radiochromatogram of an ether extract of microsomes incubated with [^{14}C]laurate. Laurate and metabolites were separated on an Altex 5 μM C_{18} (4.6 × 250 mm) column using a gradient of acetonitrile and 1% acetic acid maintained at a flow rate of 1.5 mL/min. HPLC scintillant to effluent ratio was 3; flow cell volume was 400 μl. (From Ref. 3, used with permission.)

tion provides several distinct advantages over conventional methodology in the conduct of these studies.

Radioisotopes are often used in pharmacokinetic studies in animals and man because the presence or absence of circulating metabolites can be documented by radiochemical separation techniques. The quantitation of circulating metabolites is especially important in cases where metabolites are pharmacologically active and plasma concentrations of unchanged drug do not correlate with expected pharmacologic effect. An example of such a study in which the pharmacokinetics of a leukotriene receptor antagonist, [^{14}C]SKF 102081, was determined in the guinea pig [9] is depicted in Figure 11. The concentration of [^{14}C]SKF 102081 in blood was determined by HPLC with radiochemical detection. The difference between total radioactivity and that observed for [^{14}C]SKF 102081 was accounted for by the ap-

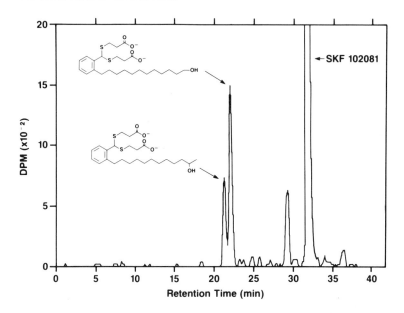

Figure 10 HPLC radiochromatogram of an ether extract of microsomes incubated with [^{14}C]SKF 102081. SKF 102081 and metabolites were separated on a Zorbax C_{18} (4.6 × 250 mm) column using a gradient of acetonitrile and 1% acetic acid maintained at a flow rate of 1.5 mL/min. HPLC effluent to scintillant ratio was 1:3; flow cell volume was 400 µl. (From Ref. 8, used with permission.)

pearance of metabolites at later time points, metabolites that could be separated from SKF 102081 by HPLC.

In mass balance/excretion studies, a large percentage, if not all, of drug excreted is in the form of metabolic products. Therefore, it is essential to separate and quantitate the individual metabolites excreted so that the relative contribution of specific pathways to the overall metabolic fate of a compound can be documented. HPLC with radiochemical detection is superior to most other techniques in this type of study because metabolites, which are often unidentified, do not have to be synthesized for quantitation. Furthermore, radiochemical techniques generally require less manipulation of samples than conventional techniques. For many metabolites, especially conjugates, sample preparation can lead to degradation. An example of the urinary metabolite profile of the lipoxygenase inhibitor, SKF 86002, as determined by HPLC with flow-through radiochemical detection, is shown in Figure 8. Unmetabolized SKF 86002, as indicated by U7 on the radiochromatogram, accounts for very little of the total radioactivity excreted in the urine.

A third area where HPLC with radiochemical detection is gaining increasing usage is in tissue distribution studies. Several investigators are now using this technique to document the exact form of the radiolabel, whether metabolites or un-

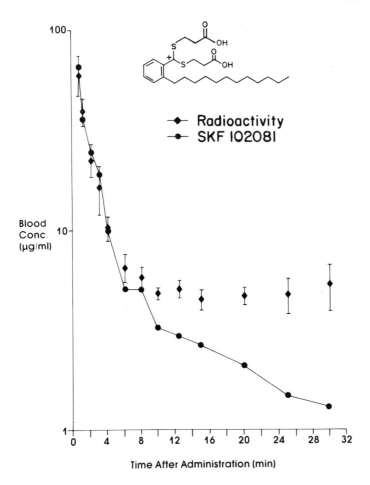

Figure 11 Disappearance of radioactivity and SKF 102081 from whole blood following administration of [^{14}C]SKF 102081. [^{14}C]SKF 102081 was administered to guinea pigs and blood was sampled at predetermined intervals. Total radioactivity was determined by LSC. SKF 102081 was quantitated by HPLC-LSC. Separation was carried out on a NOVAPAK C_{18} (8 × 100 mm) column using a gradient of acetonitrile and 1% acetic acid maintained at a flow rate of 2.5 mL/min. Column effluent was collected directly into minivials in 0.5 min fractions and then analyzed by LSC. Values for radioactivity are the mean ± SEM of at least four animals; values for SKF 102081 are from a single animal. (From Ref. 9, used with permission.)

changed drug, in specific tissues [10]. Hammerstrom and coworkers [11] have utilized HPLC-LSC to document the presence of metabolites of leukotriene C_3 in certain tissues following administration of leukotriene C_3 to mice. Murphy and coworkers [12] have also used HPLC-LSC to document the pulmonary metabolism of leukotriene C_4, and retention of metabolic products within the lung, following intratracheal administration of leukotriene C_4 to isolated perfused lungs.

In summary, HPLC coupled with radiochemical detection, either flow-through or via fraction collection, provides unique advantages in many types of metabolism studies over conventional methodologies. The selectivity, sensitivity, and reduction of sample manipulation that radiochemical technology provides often lead to superior studies in terms of design and conduct as well as interpretation.

ACKNOWLEDGMENTS

The authors would like to thank Seve Wunderly, Denis Keohane, and Lee Ann Yodis, Jeanie Eckardt, and Maria Romano for helpful suggestions and access to unpublished data and Barbara McLaughlin and Marge Schnellen for assistance in manuscript preparation.

REFERENCES

1. T. R. Roberts, Radiochromatography, *Journal of Chromatography Library*, *14* (1978).
2. G. B. Sieswerda, H. Poppe, and J. F. K. Huber, Flow versus batch detection of radioactivity in column liquid chromatography, *Anal. Chem. Acta*, *78*: 343 (1975).
3. M. C. Romano, K. M. Straub, R. D. Eckardt, L. Yodis, and J. F. Newton, Determination of lauric acid hydroxylase activity by HPLC with flow thru radiochemical quantitation, *Anal. Biochem.*, *170*, 83–93 (1988).
4. M. J. Kessler, Quantitation of radiolabeled compounds eluting from the HPLC system, *J. Chrom. Sci.*, *20*: 523 (1982).
5. T. C. Orton and G. L. Parker, The effect of hypolipidaemic agents on the hepatic microsomal drug metabolizing enzyme system of the rat, *Drug Metab. Disp.*, *10*: 110 (1982).
6. D. E. Williams, S. E. Hale, R. T. Okita, and B. S. S. Masters, A prostaglandin ω-hydrolase cytochrome P-450 purified from lungs of pregnant rabbits, *J. Biol. Chem.*, *259*: 4600–14608 (1984).
7. P. R. Ortiz de Montellano and N. O. Reich, Specific inactivation of hepatic fatty acid hydroxylases by acetylenic fatty acids, *J. Biol. Chem.*, *259*: 4136–4141 (1984).
8. J. F. Newton, K. M. Straub, R. H. Dewey, T. B. Leonard, C. D. Perchonock, M. E. McCarthy, J. G. Gleason, and R. D. Eckardt, In vitro omega hydroxylation of the leukotriene receptor antagonist 5-(2-dodecylphenyl)-4,6-dithianonanedioic acid (SKF 102081), *Drug Metab. Disp.*, *15*: 161 (1986).
9. J. F. Newton, K. M. Straub, G. Y. Kuo, C. D. Perchonock, M. E. McCarthy, J. G. Gleason, and R. K. Lynn, In vivo metabolism of the leukotriene receptor antagonist 5-(2-dodecylphenyl)-4,6-dithianonanedioic acid (SKF 102081) in the guinea pig, *Drug Metab. Disp.*, *15*: 168 (1986).

10. K.-O. Vollmer, W. Klemisch, and A. von Hodenberg, High performance liquid chromatography coupled with radioactivity detection: A powerful tool for determining drug metabolite profiles in biological fluids, *Z. Natürforsch., 41c*: 115 (1986).
11. L.-E. Apppelgren and S. Hammerstrom, Distribution and metabolism of ^{3}H-labeled leukotriene C_3 in the mouse, *J. Biol. Chem., 257*: 531 (1982).
12. T. W. Harper, J. Y. Wescott, N. Voelkel, and R. C. Murphy, Metabolism of leukotrienes B_4 and C_4 in the isolated perfused rat lung, *J. Biol. Chem., 259*: 14437 (1984).

6

HPLC with Computerized Diode Array Detection in Pharmaceutical Research

Ludwig Huber

Hewlett-Packard GmbH, Walbronn, Federal Republic of Germany

H. P. Fiedler

University of Tübingen, Tübingen, Federal Republic of Germany

INTRODUCTION

The most widely used detectors for HPLC are based on the measurement of the ultraviolet/visible (UV/VIS) absorbance properties of the analytes. These detectors are stable, selective, and sensitive, with linearity over more than 2 absorbance units. They can also be used in gradient elution analyses. These characteristics have made the UV/VIS detectors fairly universal in HPLC analysis.

The first UV/VIS detectors were fixed wavelength detectors using light sources with narrow emission characteristics and optical filters. Since the absorption of an analyte is dependent on its chemical structure and functional group(s), the versatility of these early detectors was much curtailed. Variable wavelength detectors were subsequently developed. By means of a filter or a grating monochromator, light beams with narrowly defined bandwidths could be selected from a polychromatic source for detection of solutes what would be otherwise detected poorly or not at all by the fixed wavelength detectors. The variable wavelength detectors significantly improve the versatility, selectivity, and sensitivity of the UV/VIS detection. The major limitation of the fixed and variable wavelength detectors is that only one particular wavelength window can be monitored at any given time for a given analysis.

A UV/VIS spectrum, which records the variation of absorbance as a function of wavelength, is valuable in providing information on the chemical functionalities of a compound. The first attempt to record the spectrum of a solute eluting off an HPLC column was accomplished by the use of a variable wavelength detector modified for scanning by the attachment of grating monochromator to an electric motor. Because the motion of the monochromator throughout the absorption region is slow relative to the time of solute elution, the flow of the mobile phase must be interrupted by stopping the pump and trapping the peak in the flow cell. This "stopped flow" approach is tedious and time-consuming. It was used only for very specific and difficult chromatographic problems.

With the development of photodiode array detectors [1-3], which use no moving scanning optics, acquisition of the spectrum of a solute can be done within a few milliseconds. The new rapid scanning photodiode array detectors with additional digital processing capabilities open up new applications in HPLC. Homogeneity of a chromatographic peak can be checked by overlaying normalized spectra from different sections of the peak. Furthermore, purity of a chromatographic peak can be tested by plotting the absorbance ratio at two characteristic wavelengths as a function of time. A plot with a straight line extension from the beginning to the end of the peak indicates homogeneity of the solute, provided that the absorption spectra of the potential interferences are significantly different. Unknown compounds can also be identified by comparing the spectra of the peaks with the standards.

With the diode array detector, detection sensitivity can be enhanced by widening the spectral bandwidth and summing several parallel signals with the same background. Since the wavelength bandwidth is variable, it can be optimized to produce highest detection sensitivity and linearity with lowest noise [3,4]. Changing of the refractive index and shifting of the baseline are minimized by an optical reference rather than a reference flow cell to provide a more stable gradient elution analysis. Because the grating in a diode array detector is stationary, reliable and reproducible wavelength measurements are ensured.

Several reports on the HPLC analysis of pharmaceuticals with diode array detection have been published. Carter et al. [5] used a diode array detector for checking the sample purity by examining the homogeneity of the chromatographic peaks. As little as 0.1% of the coincident impurities could be detected in the drug samples. Clark et al. [6] used new computer techniques for the analysis and characterization of Zimeldine, an antidepressant, and its primary metabolites with a diode array detector. Techniques including pseudoisomeric graphic routines, contour maps, absorbance ratios, and second derivative transformations of the elution profiles were used to manipulate the peaks for checking purity. They also reported optimization of the detector sensitivity by varying the wavelength bandwidth. Detection of Norzimeldine was at least six times more sensitive than with a regular single channel detector. The advantages of HPLC screening of drug metabolites [7] and microbial fermentation products [8,9] with diode array detection were all reported.

In the following discussion, we shall give a short overview of the principle and instrumentation of the diode array detector and its application in pharmaceutical research. The use of several computer algorithms for checking peak uniformity and identifying compounds will be demonstrated.

PRINCIPLE AND INSTRUMENTATION OF DIODE ARRAY DETECTION IN HPLC

In the conventional variable wavelength detector, polychromatic light is sent to a monochromator that generally comprises mirrors, slits, and a grating. The monochromator disperses the light so that only a narrow beam passes through the flow cell onto the photomultiplier or photodiode.

A simplified diagram of the optical design of the HP1040 diode array detector is shown (Figure 1). Polychromatic light from the deuterium lamp is precisely focused onto the window of the flow cell by an achromatic lens system. The light exiting the flow cell is dispersed by a diffraction grating, by what is known as the reversed optics, onto the photodiode array, which consists of 211 diode elements, each of 50 µm wide.

At the start of each scan, the capacitor with its associated diode is discharged by photocurrent when the diode is illuminated. The intensity of light falling onto the diode is proportional to the amount of current needed to recharge the capacitor as the diode is read out by the computer. A spectrum of the analyte from 190–600 nm can be scanned in 10 msec. Within this time, all 211 photodiodes are read out, generating more than 20,000 data points per second. Theses data must be processed by the computer at similar speed before the acquisition of next spectrum. During data processing, dark current correction, zero offset, gain adjustment, and logarithmic conversion take place.

The diode array detector becomes more valuable when the enormous amount of information generated is fed into a fast color graphic computer that processes, organizes, and stores the spectral data. For our experiments, we used an HP1040M diode array detector system from Hewlett-Packard (Figure 2). The computer is

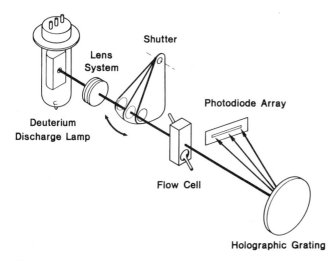

Figure 1 Schematic of a UV/VIS diode array detector.

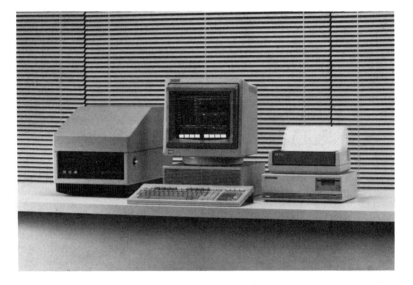

Figure 2 The HP1040M diode array detection system.

based upon the HP9000 series 300, with a 1 Mbyte of RAM and a 68010 microprocessor from Motorola. Programs, methods, and data files are stored on a Winchester disc ranging from 10 to 400 Mbyte. Large disc capacity is essential when considering the acquisition of data from a long analysis over an extended period of time. Chromatograms, spectra, and reports are printed on a Thinkjet printer or a multi-color plotter for publication purposes. A color monitor can also be used to facilitate the presentation of multisignal chromatograms, contour maps, and overlay spectra.

SCREENING FOR NEW MICROBIAL PRODUCTS

Various antibiotics including penicillins, tetracyclines, streptomycins, and chloramphenicols are products of bacterial fermentation. The author wanted to monitor the kinetics of microbacterial production of these antibiotics in the fermentation processes. To accomplish this, methods for the isolation, purification, and quality control of the active products were needed. HPLC has proved to be the method of choice for the semipreparative and analytical separation of these antibiotics [8,9]. HPLC with diode array detection is now applied for the identification and classification of the parent compounds.

Analysis of Nikkomycins

Nikkomycins produced by different strains of *Streptomyces tendae* Tü 901 belongs to the group of nucleoside–peptide antibiotics [8]. These antibiotics are classified into two groups, one containing a pyrimidine ring system, e.g., nikkomycins C_z, J,

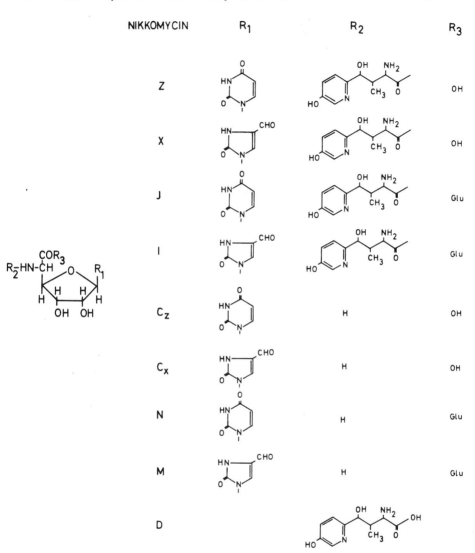

Figure 3 Structures of nikkomycins produced by *Streptomyces tendae*. (Reprinted with permission from Ref. 8.)

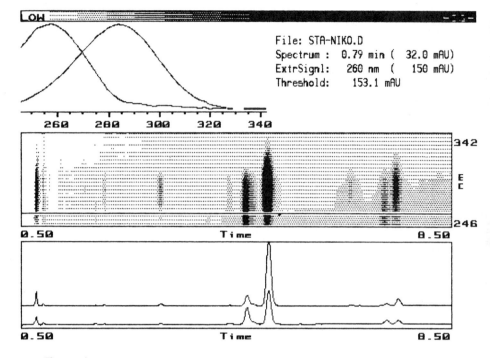

Figure 4 Display of chromatogram, contour map, and spectra of nikkomycins.

N, and Z, the other containing an imidazole ring system, e.g., nikkomycins C_x, I, M, and X (Figure 3).

Fast Screening of the Standard for Purity

In classifying the Nikkomycins, or any new compound for that matter, the problem most encountered is the unavailability of reference standards. In this study, standards were obtained by collecting fractions containing the major products from the ion-exchange separation of the culture filtrate. The elute fractions were subsequently lipophylized and reinjected into the HPLC, and the spectra of the standards were scanned with the diode array detector and stored on the Winchester disc at the frequency of 2 Hz. The computer-evaluated data were displayed on the screen as shown (Figure 4). From the lower window up are the chromatogram, the contour map, and the spectra of the peaks.

The chromatogram is constructed from the stored spectra so that the signal at the wavelength of maximum absorbance of the compound is displayed. This guarantees that any compound with the absorbance spectrum above the threshold will be represented on the chromatogram.

The contour map is a three-dimensional presentation of the diode array detector scans. The data set is viewed from above the wavelength/time plane. The x axis is time as in the chromatogram, the y axis is the wavelength, and the z axis, the intensity of absorbance, which appears in different colors with a colored monitor and in different gray shades with a monochrome monitor.

On screen, the wavelength of the spectrum and the elution time of the peak are selected by softkeys. When the vertical cursor is invoked for time selection and placed on a chromatographic peak, a normalized spectrum corresponding to the time will be displayed in the upper left window. The digital values of the wavelength cursor position and the time cursor positions will be displayed in the upper right window.

For checking of peak purity, the spectra at various times along the peak are compared. The spectrum at up slope or down slope of a peak is selected by placing the time cursor at the respective positions on the peak. The spectrum is displayed in the upper left window. The cursor is then set free and moved to the apex of the peak. If significantly difference is found in the overlay of the spectrum at the apex of the peak, it can be assumed that the chromatographic peak contains impurity. With this approach, it was found that the first peak at 0.79 min in Figure 4 consisted of at least two components.

Optimization of the Wavelength for the Entire Analysis

For optimizing the wavelength of observation for the chromatogram, a horizontal cursor associated with the contour map is used. A second chromatogram will be displayed below the first at the wavelength selected by the horizontal cursor in the contour map (Figure 4). When the horizontal cursor is moved along the wavelength axis in the contour map, the intensity of the peaks in the second chromatogram changes simultaneously. In this way, the wavelengths of observation for the second chromatogram can be optimized for the imidazole type nikkomycines at λ_{max} 290 nm or the pyrimidine type at λ_{max} 260 nm (Figure 4).

Optimization of the Wavelength Bandwidth for Individual Peaks

If maximum detection is required, a wavelength bandwidth equal to the width at half height of the analyte spectra has been found to be optimal [3]. However, in most applications, selectivity was more important than sensitivity, and a practical bandwidth of 4nm was chosen for these applications.

Quantitation of Coeluting Compounds with the Spectral Suppression Technique

As shown in Figure 4, the first peak with retention time of 0.79 min contained two unresolved compounds. Because these two compounds have significantly different properties, they can be quantified by the spectral suppression technique as described in the work of Fell et al. [10]. By this technique, the interference is subtracted from the chromatogram if equiabsorptive wavelengths are used for the signal and background wavelength.

The spectra of nikkomycin C_x with λ_{max} at 290 nm and an unknown with λ_{max} at 260 nm are shown (Figure 5). For quantitation of nikkomycin C_x, a signal wave-

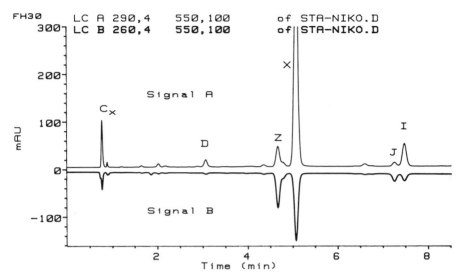

Figure 5 Chromatograms of nikkomycin standards display at 290 and 260 Column: ODS Hypersil; 125 x 4.6 mm. Solvent A: H_2O—0.01 M heptane sulf. acid—0.2% acetic acid. Solvent B: H_2O/ACN 6/4—0.01 M heptane sulf. acid—0.2% acetic acid. Gradient: from 13 to 40% B in 9 min.

length above 290 nm can be used. Above this wavelength the unknown antibiotic has low UV absorbance and did not interfere with nikkomycin detection (Figure 6).

The unknown antibiotic is best detected at 260 nm which is the wavelength of maximum absorbance. To quantify the unknown antibiotic, the signal of nikkomycin C_x at 260 nm must be subtracted from the signal of the unknown. The point with equiabsorbance to 260 nm for nikkomycin C_x was found to be 308nm. The unknown antibiotic was then detected at 260 nm with the background at 308 nm subtracted (Figure 7).

The spectral suppression technique is useful only when the spectrum of a compound is well defined. In the unfortunate cases where the spectra are similar, the technique can still be applied [10], but the gain in selectivity is at the expense of sensitivity.

Signal and Spectral Comparison of Standards and Antibiotic Products in the Culture Broth

The chromatograms shown in Figure 8 reveal that several peaks found in the chromatogram of the culture filtrate had retention times that coincided with the chromatogram of the standards. The identity of the peaks at 3.0 and 7.2 min were further confirmed by overlaying the antibiotic spectra with the standard spectra.

HPLC with Computerized Diode Array Detection

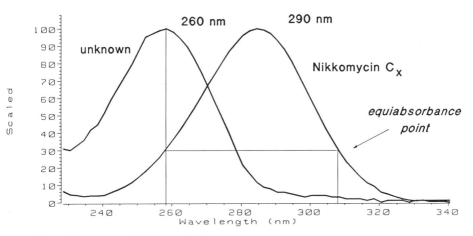

Figure 6 Spectra of nikkomycin C_x.

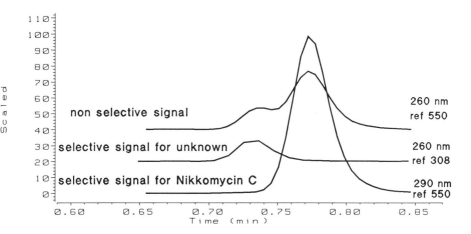

Figure 7 Excerpt from chromatogram in Figure 4 showing multiple signal detection and peak suppression. At 290 nm the unknown, which does not absorb UV, is not detected (below). At 260 nm with a reference of 550 nm both the unknown and nikkomycin C are detected (upper). At 260 nm with a reference of 308 nm, the spectral contribution of nikkomycin C is suppressed and the unknown alone is detected (middle).

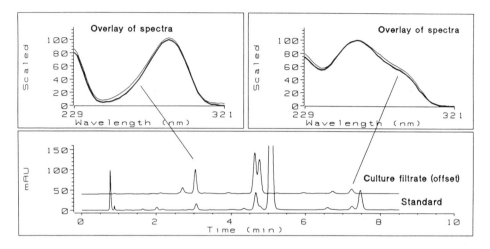

Figure 8 Display of overlay spectra and chromatograms of standards, and antibiotics found in the culture broth. Conditions as in Figure 5.

Use of Absorbance Ratios for Screening Different Types of Nikkomycins

When grown under controlled conditions, nikkomycin bacteria should produce only the pyrimidine type antibiotics. The authors wanted to know how specific was this condition. There are several ways that the specificity can be checked. One way is to review all the individual spectra under the peak, and the other, a faster and more efficient way, is to review the ratiograms. Figure 9 shows the standard chromatogram at 290/260 nm for pyrimidine type nikkomycins is about 0.5 and for imidazolone type 1.5. Figure 10 shows the chromatogram from analyzing the product of fermentation along with the ratiogram. The peaks with ratio above 0.5 at 4.8 and 6.7 in were evidently new nikkomycins and were isolated for structure elucidation.

Analysis of Urdamycins

Urdamycins (Figure 11) are new angucycline antibiotics produced by *Streptomyces fradiae*. They are biologically active against gram-positive bacteria and stem cells or murine L-1210 leukemia. With *Streptomyces fradiae* we have a good example of the ability of microorganisms to produce secondary metabolites of related chemical structure in different proportions depending on the culture conditions [11].

The chromatogram of urdamycin standards is shown along with the spectra (Figure 12). The spectra show the characteristic absorption of the chinoide system between 350 and 550 nm.

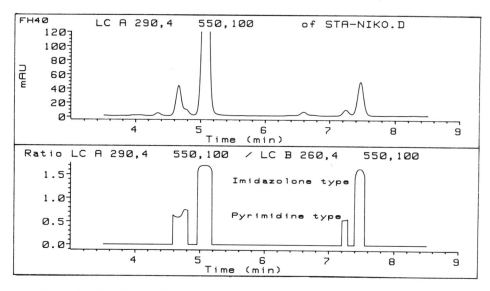

Figure 9 Absorbance ratiogram and chromatogram of nikkomycin standards. Conditions as in Figure 5.

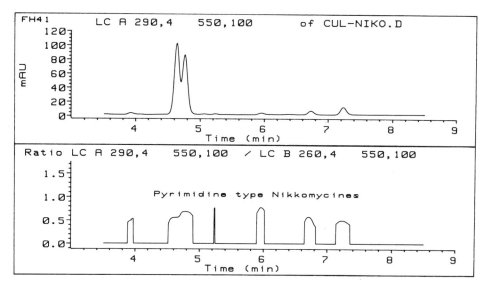

Figure 10 Absorbance ratiogram and chromatogram of nikkomycins found in culture broth. Conditions as in Figure 5.

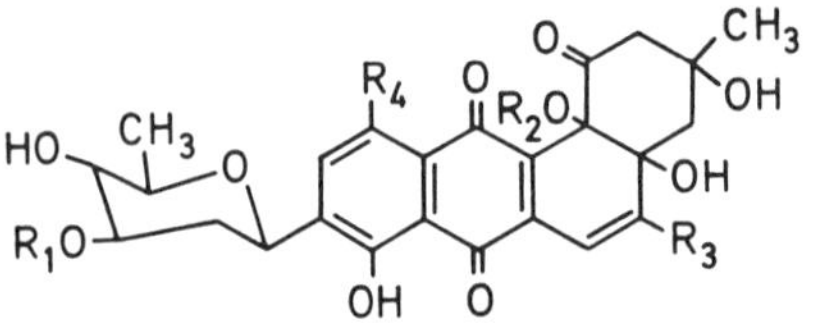

Figure 11 Basic structure of urdamycins A, C, D, E.

Peak Purity Check by Spectral and Chromatogram Overlay

Several procedures have been described for checking peak purity [5,6,10]. The most popular one is the overlaying of normal UV/VIS spectra at various sections of the peak. The alternative is the overlaying of a normalized chromatogram at different wavelengths.

The overlay of the chromatogram of urdamycin standards with the chromatogram of the antibiotic products found in the culture filtrate is shown (Figure 13). The chromatogram of the analysis in question is displayed in the lower window, the spectra at various sections of the peak on the upper left, and the overlay chromatograms on the upper right (Figure 14). From the normalized spectra and chromatogram analysis, the peak at 9.3 min is impure. The overlay chromatograms show three unresolved peaks, with two components in the major peak (Figure 14, upper right).

Peak Purity Check by Absorbance Ratios

Drouen et al. [12] formulated the guidelines for using dual wavelength absorbance ratios for solute purity check. They found that the influence of the background of mobile phase on the UV spectra was probably the major limitation in the practical application of the technique. By plotting the absorbance ratio of two signals at different wavelengths, interference is suspected if the ratio changes across the peak. However, in plotting the peak absorbance ratio, care must be taken in selecting the appropriate noise threshold. If the threshold is too low, the ratio will merely reflect noise fluctuation. If the threshold is too high, some of the detection signal will be lost, and no ratiogram will be generated for the beginning and end of the peak. A portion of the chromatogram and the associated ratiogram generated by dividing the signal at 320 nm by 225 nm are shown (Figure 15). With a detection threshold of 1, the ratiogram showed peaks at 9.2 and 9.5 min to be impure. With a detection threshold of 10, it was still apparent that the peak at 9.2 was impure, but the impurity in the second peak was undetected. With a threshold of 18, the detection sensitivity deteriorated and the peak width in the ratiogram reduced.

HPLC with Computerized Diode Array Detection

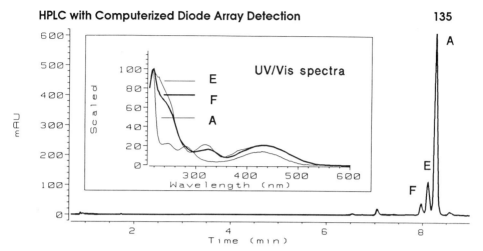

Figure 12 Chromatogram and the UV/VIS spectra of urdamycin standards. Column: ODS Hypersil; 5 µm, 125 x 4.6 mm Gradient: 0.1% phosphoric acid/acetonitrile from 0 to 60% within 12 min; flow rate: 2 mL/min.

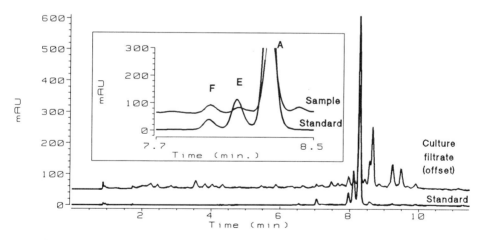

Figure 13 Display showing the overlay of the chromatogram of products found in the culture broth with the chromatogram of the urdamycin standards. Conditions as in Figure 12.

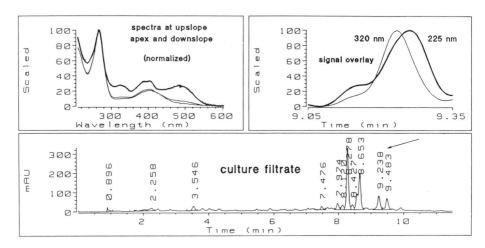

Figure 14 Peak purity check by overlaying of spectra and zoomed chromatograms. Conditions as in Figure 12.

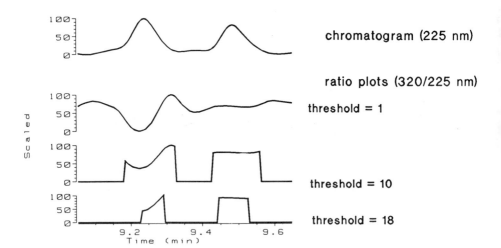

Figure 15 Chromatogram and absorbance ratiograms with different thresholds. Conditions as in Figure 12.

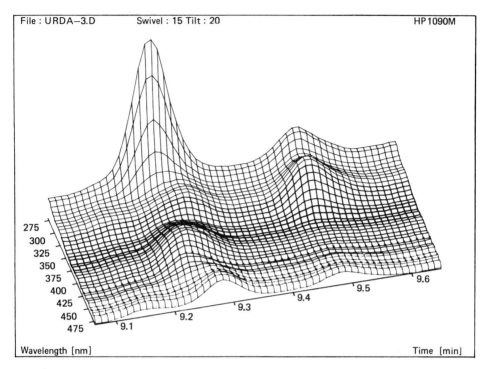

Figure 16 The three-dimensional spectro-chromatograms of urdamycins.

Peak Purity Check by Three-Dimensional Spectrochromatogram

The three-dimensional spectrochromatogram is another alternative for fast examination of peak purity (Figure 16). The plot shows clearly that the peak maxima at 9.2-9.3 min varied as a function of the wavelengths. The peak at this retention time consists of at least two compounds.

Comparison of Products Found in the Culture Broth and the Cell Extract

The authors wanted to know whether the products in the culture filtrate were qualitatively the same as those found in the methanol extract from the bacterial cell. The chromatogram of the overlay of the cell extract with the culture filtrate is shown (Figure 17). It shows that the early eluting peaks in the chromatogram of the cell extract are somewhat higher than in the culture filtrate. The authors wanted to see whether these early eluting peaks were indeed urdamycins. This could be done rather quickly with the aid of a three-dimensional spectrochromatogram (Figure

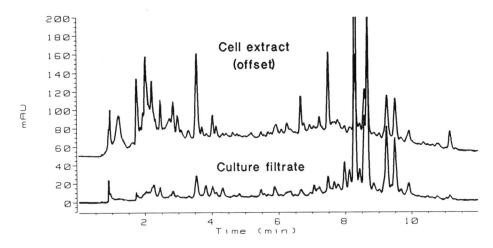

Figure 17 Chromatograms of compounds found in the culture broth and in cell extract of urdamycin bacteria. Conditions as in Figure 12.

18). These early eluting compounds do not absorb above 350 nm as do urdamycins, and therefore they could not be such, and were not further investigated.

Screening for Urdamycins with a Maximum Absorbance Signal Chromatogram at a Selected Wavelength Range

The example just shown offers a fast overview of the spectral characteristics of the sample for urdamycin identification and peak homogeneity by the more intense peaks in the plot. To circumvent this, a maximum absorbance signal chromatogram at a selected wavelength range may be used for screening unknown urdamycins.

Since various urdamycins exhibit absorption maxima between 360 and 600 nm, there is no single optimal wavelength for reconstructing the chromatogram. For producing the maximum absorbance signal plot, a routine was used that searches the stored spectrum in a specified wavelength range for the maximum absorbance at a given time. The wavelength range 210-600 nm was chosen for the universal signal, and 360-600 nm for urdamycins (Figure 19). Without knowing the absorbance maxima of the compounds separated in the chromatogram, this routine always plots the signal at the maximum absorbance within a specified wavelength range. By supplying a wavelength range specific to the compounds of interest, the specific detection of the compounds is facilitated. Eight compounds with wavelength range from 300-600 nm were found to elute in the chromatogram. By spectrocomparison, these compounds were confirmed to be new urdamycins. The advantage of the maximum absorbance signal plot over the three-dimensional plot is that the signal from the

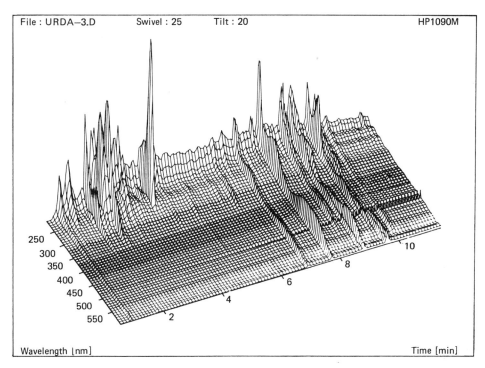

Figure 18 Three-dimensional spectro-chromatograms of a cell extract showing the late elution of urdamycins.

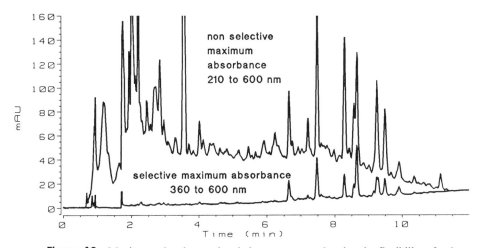

Figure 19 Maximum absorbance signal chromatograms showing the flexibility of universal as well as specific detection of the diode array detector. Conditions as in Figure 12.

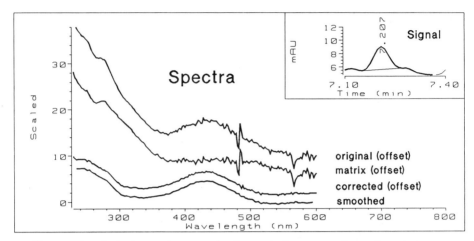

Figure 20 The background-corrected and smoothed spectrum of an urdamycin.

former can be integrated in the usual manner, and the minor peaks are better identified.

Noise Reduction of a Spectrum

When an urdamycine is present in a very small concentration, the recorded absorbance spectrum will be noisy (Figure 20). The first approach to reduce the noise and obtain a corrected spectrum is to subtract a background spectrum taken at the baseline followed after the peak. The noise of the corrected spectrum can be further reduced by the signal smoothing algorithms described by Sawitzky et al. [13].

PHARMACOKINETICS AND DRUG METABOLISM

The absorption spectra of metabolites often resemble that of the parent drug. This similarity is the key to success in using diode array detectors in pharmacokinetics study of drugs. Thus metabolites are identified by comparing spectra of the unknown peaks with those of the parent drug. Similarity of the unknown spectra to the standard spectra indicates the presence of a potential metabolite, and it calls for trapping the peak for further investigation.

Identification of Urapidil Metabolites

In studying the pharmacokinetics and metabolism of a drug, the metabolites must be recovered and analyzed from some kind of physiological fluid, such as urine, or blood. Typically, these matrices are loaded with endogenous metabolites that may coelute and interfere with the identification of the drug. The use of the diode array detector for differentiating urapidil and its metabolites (Figure 21) from the endogenous interferences has been reported [7]. Metabolite candidates were selected

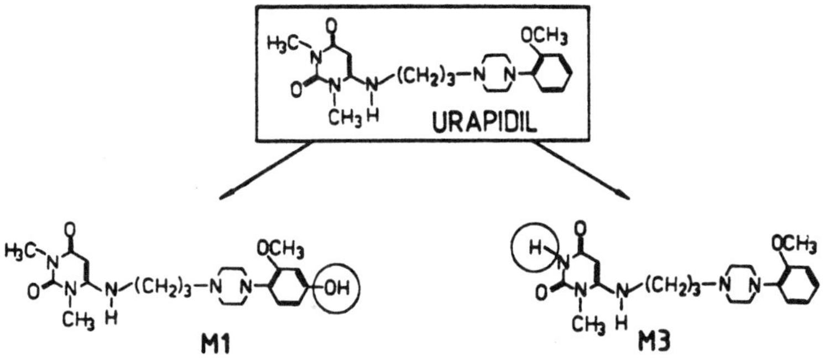

HPLC with Computerized Diode Array Detection

Figure 21 Structure of urapidil and two metabolites. (Reprinted with permission from Ref. 7.)

based on absorbance ratios. The overlay of the chromatograms at 230 nm (solid line) and at 268 nm (dotted line) of the urine of a dog that had been administered urapidil is shown (Figure 22). Urapidil absorbs with a maximum at 268 nm and a minimum at 230 nm (Figure 23). Therefore, a metabolite is suspected whenever a peak is taller in the dotted line than in the solid line. Peaks MX, M1, and M3 were identified as metabolites by spectral matching with the urapidil standard (Figure 23).

Peak Homogeneity and Identity of Theophylline in Serum

Theophylline in serum was measured by HPLC with diode array detection [14]. An aliquot of the serum sample was injected directly onto a cleanup column. Theophylline from the cleanup column was introduced to the analytical column through a switch valve. The chromatogram of theophylline with the associated spectra is shown (Figure 24). The upper windows of the display from left to right are the superimposed spectra from the various sections of the theophylline peak for checking purity and the overlay spectra of the theophylline peak with the standard for confirming identity.

IDENTIFICATION OF PHARMACOLOGICALLY ACTIVE NATURAL PRODUCTS

HPLC with diode array detection in combination with a library of spectra has been used to identify natural products. A number of methods have been proposed for archival retrieval of spectroscopic data [15] and UV/VIS diode array spectra [4].

Schuster [16] reported the analysis of flavonoids from plant extracts. The chromatogram of the plant extract with the associated spectra is shown (Figure 25). The user can select any peak in the chromatogram with a cursor for identification. As shown in the chromatogram, the identity of the peak marked with a ? was desired. The library of spectra in the computer was searched fro spectra that matched the background corrected spectrum from the unknown peak. When a match was found,

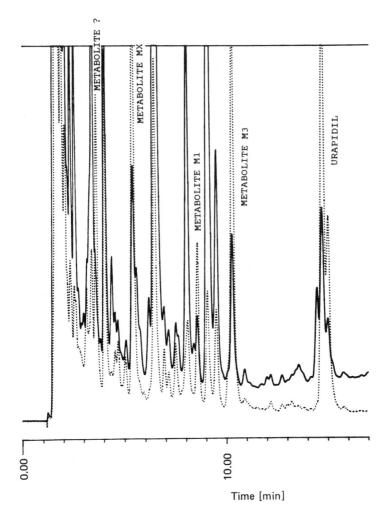

Figure 22 Chromatograms of urapidil and its metabolites plotted at 230 nm (solid line) and 268 nm (dotted line). Column: ODS Nucleosil, 125 × 4.6 mm. Gradient: 0.05 M sodium perchlorate, from 0 to 30% methanol within 13 min; flow rate: 1.0 mL/min. (Reprinted with permission from Ref. 9.)

HPLC with Computerized Diode Array Detection

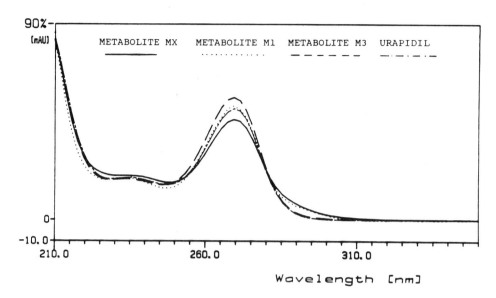

Figure 23 Spectra of urapidil and metabolites. (Reprinted with permission from Ref. 7.)

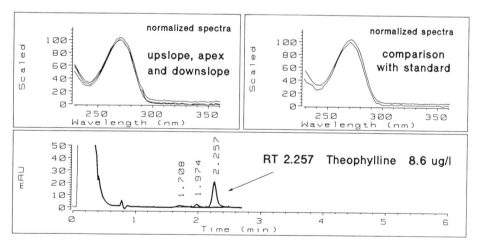

Figure 24 Chromatogram and overlay spectra of theophylline in a serum sample. Column: ODS Hypersil; 100 x 2.1 mm, 5 μm. Gradient: 0.01 M KH_2PO_4, from 0 to 25% acetonitrile within 3 min; flow rate: 1 mL/min.

the fitness of the match was automatically calculated by least square regression. The normalized spectra of the two best candidates were then overlaid, and the names of the compounds and the correlation factors were displayed (Figure 26). As suggested by visual comparison of the spectra as well as the correlation factors, the unknown was identified as hyperoside.

Spectra libraries can be used for interactive and automated peak identification. Retention time pre-search is recommended because it reduces library search time and improves the reliability of the search result. In such cases, the spectra library can be used for one set of chromatographic conditions only and is part of the analysis method like integration parameters or a calibration table.

CONCLUSION

HPLC with UV/VIS diode array detection and fast computer graphics is an invaluable tool for pharmaceutical research. The diode array detector helps to optimize the wavelengths for more sensitive detection, more specific identification, and more accurate quantification, as demonstrated in the isolation of various types of nikkomycins.

Absorbance ratio plots and three-dimensional spectrochromatograms are useful for screening products of microbial fermentation in culture broths and for products of drug metabolism in physiological fluids. HPLC with diode array detection in combination with a library spectra has also been used for identification of natural products.

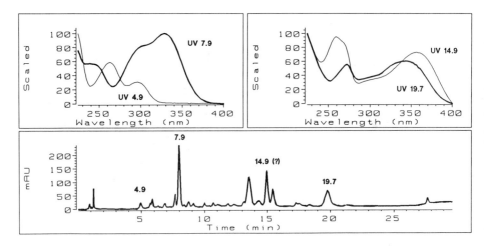

Figure 25 Chromatogram, and the associated spectra of a plant extract. Column: ODS Hypersil; 100 x 4.6 mm; 5 µm. Gradient: 0.005 M KH_2PO_4, from 5 to 70% methanol within 30 min.

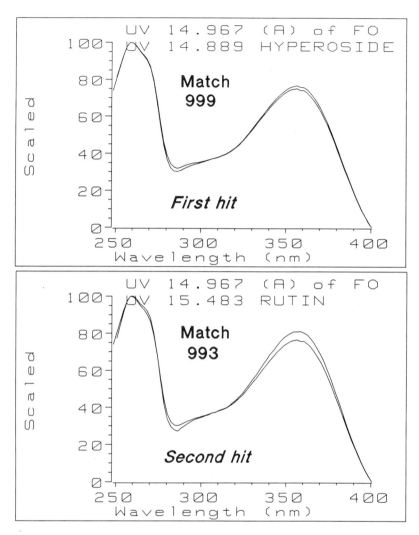

Figure 26 Overlay spectra showing the two best possibilities of an unknown. Visual inspection as well as the correlation factors suggest hyperoside to be the compound.

REFERENCES

1. M. J. Milano, S. Lam, and E. Grushka, Rapid scanning diode array as a multiwavelength detector in liquid chromatography, *J. Chromatogr.*, *125* 315-326 (1976).
2. S. A. George and A. Maute, A photodiode array detection system: Design concept and implementation, *Chromatographia*, 419-425 (1982).
3. E. Elgass, A. Maute, R. Martin, and S. George, A high-speed spectrophotometric detector for HPLC, *American Laboratory*, Sept. (1983).
4. A. F. Fell, B. J. Clark, and H. P. Scott, Computer-aided strategies for archive retrieval and sensitivity enhancement in the identification of drugs by photodiode array detection in high-performance liquid chromatography, *J. Chromatogr.*, *316*: 423-440 (1980).
5. G. T. Carter, R. E. Schiesswohl, H. Burkle, and R. Yang, Peak homogeneity determination for the validation of high-performance liquid chromatography assay methods, *J. Pharm. Sci.*, *71*: 317-321 (1982).
6. B. J. Clark, A. F. Fell, and H. P. Scott, Rapid-scanning, multi-channel high-performance liquid chromatographic detection of zimeldine and metabolites with three-dimensional graphics and counter plotting, *J. Chromatogr.*, *286*: 261-273 (1984).
7. K. Zech, R. Huber, and H. Elgass, On-line screening for drug metabolites by high-performance liquid chromatography with a diode array UV detector, *J. Chromatogr.*, *282*: 161-167 (1983).
8. H. P. Fiedler, Screening for new microbial products by high-performance liquid chromatography using a photodiode array detector, *J. Chromatogr.*, *316*: 487-494 (1984).
9. H. P. Fiedler, Identification of new elloramycines, anthracycline-like antibiotics in biological cultures by HPLC and diode array detection *J. Chromatogr.*, *361*: 432-436 (1986).
10. A. F. Fell, H. P. Scott, R. Gill, and A. C. Moffat, Novel techniques for the peak recognition and deconvolution by computer-aided photodiode array detection in high-performance liquid chromatography, *J. Chromatogr.*, *282*: 123- 140 (1983).
11. H. Drautz, H. Zaehner, J. Rohr, and A, Zeeck, Urdamycines, new angucycline antibiotics from streptomyces fradiae, *J. Antibiot.*, in press.
12. A. J. H. Drouen, H. A. H. Billiet, and De Galan, Dual-wavelength absorbance ratio for solute recognition in liquid chromatography, *Anal. Chem.*, *56*: 971-978 (1984).
13. A. Sawitzky and M. J. E. Golay, Smoothing and differentiation of data by simplified least square procedures, *Anal. Chem.*, *36*: 1627-1639 (1984).
14. M. Riedmann and L. Huber, Automated serum samples analysis using sample cleanup columns in LDLC, *Amer. Clin. Products*: 8-15, April (1986).
15. J. Zapan, M. Penca, D. Hadzi, and J. Marsel, Combined retrieval system for infrared, mass, and carbon-13 nuclear magnetic resonance spectra, *Anal. Chem.*, *49*: 2141-2146 (1977).
16. R. Schuster, HPLC determination of flavonoids and phenolic compounds in natural products, Hewlett-Packard Application Brief, HP P/N 12-5953-0081, March (1983).

7
Design and Application of HPLC/FT-IR

Kathryn S. Kalasinsky*

Mississippi State Chemical Laboratory, Mississippi State, Mississippi

Victor F. Kalasinsky*

Mississippi State University, Mississippi State, Mississippi

INTRODUCTION

In recent years, considerable effort has gone into coupling chromatographic separations with spectrometric methods that can be useful in identifying eluates. Interfaces for combining high performance liquid chromatography (HPLC) to infrared(IR) spectroscopy [1-26] and nuclear magnetic resonance (NMR) spectroscopy [27,28] represent relatively new fields of study that have not reached the stages of development that gas chromatography/mass spectrometry (GC/MS) [29], GC/IR [30,31], or HPLC/MS [32-35] have. Commercial systems are available for the latter three techniques, as is a thin-layer chromatography/infrared (TLC/IR) accessory [36], but there is very little choice among commercial systems for HPLC/IR or HPLC/NMR. With the volume of work that is currently underway, however, it is expected that the final designs will be available in the marketplace within the next few years. In this article we will briefly review progress in the area of HPLC/IR and cite representative articles. A more complete discussion will be given for examples that demonstrate an interface developed in our laboratories.

Two basic approaches have been taken for HPLC/IR: the use of flow-through infrared cells and the use of solvent elimination techniques. The most significant difficulty in combining HPLC and IR spectrometry is that the mobile phase will

---

*Current affiliation: Armed Forces Institute of Pathology, Washington, D.C.

generally have a complicated infrared spectrum of its own. Because of the large relative concentration of mobile phase, solvent interference is a distinct disadvantage of flow-cell HPLC/IR. In fact, it sometimes requires that the solvents be selected in order to optimize the infrared results rather than to optimize the chromatographic conditions. Aqueous solvent systems create the worst possible problems for flow-cell HPLC/IR since water is a very strong infrared absorber and can blank out large portions of the infrared spectrum. Vidrine has discussed the practical aspects of choosing solvents [1]. Nonetheless, useful HPLC/IR data have been obtained by using flow cells with Fourier transform infrared (FTIR) instruments.

Two types of flow cells have been used for HPLC/IR: those involving infrared transmission and those involving reflectance techniques. Transmission cells were used first and have evolved from modified cavity cells with salt windows [1-5] to specially designed low-dead-volume cells matched to chromatographic conditions [6-9]. Normal-phase solvents were utilized in the early studies, and, with careful selection of common HPLC solvents, infrared data over narrow frequency regions were obtained. Hydrocarbon solvents, however, made it very difficult indeed to observe the C-H stretches of the solutes. With the development of microbore HPLC columns, it was possible to take advantage of the reduced solvent volumes and use (more expensive) deuterated solvents to open up the very important C-H stretching region [6]. For aqueous reversed-phase chromatography, Jinno et al. [7] used a PTFE cell and a CD_3CN/D_2O solvent system with a microbore column to observe carbonyl and C-H stretches. Taylor and coworkers [8] have offered two solutions to the problems of reversed-phase flow-cell HPLC/FT-IR. They constructed a low-dead-volume cell from CaF_2 for microbore studies and, for analytical columns, a membrane separator in which the solute was extracted from the aqueous phase into an organic solvent that diffused through the hydrophobic membrane and was eventually sent to the infrared flow cell [9].

A flow cell that uses a cylindrical ZnSe internal reflection crystal has been designed and is commercially available [10]. The cell has a relatively large volume for HPLC work (24 µL), but reasonable results have been obtained with many solvents (including water) through the use of absorbance subtraction techniques and spectral blanking for solute concentrations of 1-2% (w/v) [11].

The second main approach to HPLC/IR involves solvent elimination followed by infrared spectroscopy. A potential disadvantage of this method relative to flow cells is that a volatile solute may be eliminated along with the solvent, but an advantage is that higher sensitivities are possible with fewer interferences. The subject of solvent-elimination HPLC/IR has been reviewed previously [12], but sufficient work has continued in this area that it is worthwhile to mention recent progress. Kuehl and Griffiths [13,14] described a system for normal-phase HPLC/IR that uses a solvent concentrator to reduce the volume of the mobile phase and a series of KCl-laden cups arranged on a carousel to hold the samples. A UV detector was used in controlling the system to ensure that solute (and not just solvent) was deposited in a sample cup, and the spectra were recorded by using the diffuse reflectance infrared Fourier

Design and Application of HPLC/FT-IR

transform (DRIFT) technique. Using microbore HPLC, Jinno deposited the column effluent onto a moving KBr plate and subsequently recorded transmission infrared spectra [15,16]. Also with a microbore system, Griffiths and coworkers [17] constructed a more compact version of their previous HPLC/FT-IR interface. Gagel and Biemann have used a reflection-absorption accessory for obtaining FT-IR spectra after depositing normal-phase microbore HPLC eluates in a spiral pattern onto the reflective substrate [18].

To address the problem of aqueous reversed-phase HPLC, Griffiths and coworkers constructed a flowing extraction technique to transfer the solutes from the aqueous medium into an organic solvent (methylene chloride), which was ultimately deposited in the sample cups of the carousel [19]. Still another approach involves the use of diamond dust in place of KCl as the substrate for diffuse reflectance. Direct elimination of water and other solvents has been demonstrated [20,21]. Jinno and coworkers have used a wire-mesh support for microbore HPLC solutes, and after allowing aqueous reversed-phase mobile phases evaporate, they recorded transmission infrared spectra [22]. Very recently, by improving their nebulizer efficiency, Gagel and Biemann were able to eliminate water and obtain reflection-absorption spectra of low nanogram quantities of eluates [23]. Still another aerosol generation interface for combining liquid mobile phases is the use of a monodisperse aerosol generation interface for combining liquid chromatography with FT-IR (MAGIC-LC/FT-IR) introduced by Browner and coworkers [24].

We recently described a system for microbore columns (normal- or reversed-phase) that uses nebulization to reduce the solvent volume and uses diffuse reflectance for infrared detection [25]. When aqueous mobile phases are present, an acid-catalyzed postcolumn reaction with 2,2-dimethoxypropane (DMP) converts water to acetone and methanol (both of which are easily volatilized) by the following scheme [37,38]:

$$H_3C-\underset{\underset{O-CH_3}{|}}{\overset{\overset{O-CH_3}{|}}{C}}-CH_3 + H_2O \xrightarrow{H^+} H_3C-\overset{\overset{O}{||}}{C}-CH_3 + 2\ CH_3OH$$

For analytical (nominal 4-mm inner diameter) columns, a splitter is used to divert a small portion of the eluent flow to the post-column reactor (if necessary) and to the infrared detection system [26]. A detailed description of the interface and examples of aqueous reversed-phase HPLC/FT-IR data are given in subsequent sections. This design has shown considerable promise and appears to be highly applicable to current pharmaceutical needs and samples.

EXPERIMENTAL PROCEDURES

Chromatographic Systems

A Waters μ-Bondapak C18 reversed-phase column (3.9 mm × 30 cm; 10 μm particles) and an Altech C18 bonded reversed-phase column (1 mm × 25 cm; 5 μm particles) were used for analytical and microbore separations, respectively. Solvents were delivered to the column and to the postcolumn mixing tee by Waters model M-45 pumps controlled by a Waters model 680 automated gradient controller. The column flow rates were generally 1.0 mL/min for analytical and 50 μL/min for microbore columns, and the controller allowed DMP flow rates to be varied between 50 and 100 μL/min as necessary. Samples were injected by using a Rheodyne 7105 (50-μL sample loop) or a model 7413 (1- or 5-μL loop) sample injection valve, depending on the interface used.

Reagents

The samples of acetaminophen, aspirin, caffeine, niacin, and nicotinamide (Aldrich Chemical Co.) were stated to be at least 98% pure. The samples of cortisone and corticosterone were purchased from Sigma Chemical Co. All samples were used without further purification. The analgesic and anticonvulsant were commercial preparations. The solvents (Burdick and Jackson Laboratories, Inc.) were all HPLC grade. 2,2-Dimethoxypropane (Aldrich, 96% pure) was slightly acidified by the addition of concentrated hydrochloric acid (12 M) resulting in a solution that had a HCl concentration of approximately 10^{-5} M.

The KCl substrate was ground to a fine powder in a motorized mortar and pestle. In order to ensure the removal of all organic impurities, the KCl was extracted with methylene chloride, dried, and heated to 500° prior to use.

Data Collection and Processing

Infrared spectra were recorded on a Nicolet 7199 Fourier transform interferometer equipped with a liquid-nitrogen-cooled MCT detector and a Harrick DRA-SN5 diffuse reflectance accessory modified to accept sample "trains" [39].

Typically, 32 scans at 4-cm^{-1} resolution were coadded and Fourier processed for each compartment of the trains. Reflectance spectra were obtained by ratioing the single-beam spectrum of each compartment to the single-beam spectrum of a compartment onto which only solvent had been sprayed. Reconstructed chromatograms were obtained by using the integrated absorbance option available with the standard Nicolet software.

INTERFACE DESIGN

A schematic diagram of the basic interface design for our aqueous reversed-phase microbore-column system is shown in Figure 1. The HPLC eluent is introduced to the 2,2-dimethoxypropane (DMP) postcolumn reactant in a mixing tee, and the combined fluids then flow through a heated (35-50°C) mixing coil to assure com-

Design and Application of HPLC/FT-IR 151

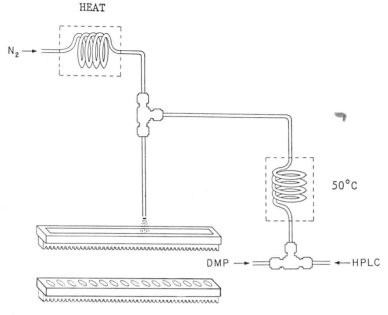

Figure 1 Schematic diagram showing the essential features of the microbore HPLC/FT-IR interfaces. Used with permission, Ref. 25.

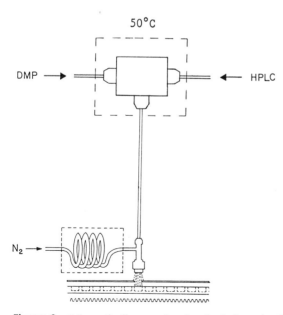

Figure 2 Schematic diagram showing the "micromixer" and AA nebulizer used in the HPLC/FT-IR interfaces.

plete reaction. The heat is necessary because the water/DMP reaction is slightly endothermic. A heated nitrogen gas stream is then introduced to a second tee to help nebulize the eluent from the 1/16" stainless steel tubing and deposit it onto the KCl substrate for analysis by DRIFT. The substrate is prepared from a methanol/KCl slurry packed into either a compartmentalized "train" or a continuous "trough." The methanol is allowed to evaporate and leave a very firm, flat surface that makes an excellent substrate for diffuse reflectance analysis [39].

The efficiency of the system can be enhanced by the use of a nebulizer from an atomic absorption (AA) burner and the addition of a more efficient, commercially available "micromixer" tee (Kratos) that eliminates dead volume. This schematic is shown in Figure 2. When the system is operated with normal-phase or nonaqueous reversed-phase solvents, it is only necessary to discontinue the DMP flow; it is not necessary to reroute any of the tubing. With the micromixer in place, the chromatographic resolution is maintained.

The sensitivity of the system was improved by approximately 10% by substituting an ultrasonic nebulizer for the AA nebulizer. This produces a finer spray that, when entrained in a low pressure nitrogen flow, facilitates evaporation of the solvent and provides more control over the spot size of the deposited sample. As the spot size (spray size) gets smaller, the sample becomes more concentrated, and the infrared detection limits are improved.

For analytical (or conventional, 3.9-mm or 4.6-mm bore) columns, where the solvent volume is much greater, a splitter is used on the flow exiting the column, and the smaller volume is sent on to the postcolumn reaction (for water elimination, if an aqueous mobile phase is used) and infrared detection. The split ratio is adjustable, but we have found that 20 to 1 works well for typical LC conditions.

The interface design can be used either on line ("on-the-flow") or off line. For on line operation, a stepper motor controls the speed of the train or trough as it is moved under the point of sample deposition. An insulated, heated chamber operated at reduced pressure is used to assist in final drying, and the sample train is stepped through the diffuse reflectance accessory in the FT-IR sampling compartment for infrared analysis.

Alternatively, the entire process of separation (HPLC), postcolumn reaction, sample deposition, and final drying can be done off line and then the train or trough brought to the infrared instrument for DRIFT analysis. This option is useful because it is sometimes inconvenient to move the entire LC apparatus to the infrared spectrometer, and once the eluates have been deposited, the entire chromatogram is "stored" on the train or trough. A chromatogram can be "reconstructed" from the infrared data after the data has been stored in a computer, thus eliminating the need for a conventional HPLC detector. With a system of this type, it is also possible to return to samples of particular interest for longer signal averaging (and better infrared data), and it is possible to recover each sample for, perhaps, further analysis by other methods.

The time resolution available for the infrared interface is also adjustable, but the length of time over which the spray is directed onto a single compartment is gener-

Design and Application of HPLC/FT-IR 153

ally between 10 and 30 sec. An obvious trade-off is between the deposition time per compartment (chromatographic resolution) and the length of the train (total chromatogram time). And while we have used long trains (up to 4') in the past, we find it much easier to use a series of shorter trains (1' long) if the chromatograms are longer than approximately 15 min. As shown in later figures, the chromatographic resolution is retained in HPLC/FT-IR experiments.

Chromatograms can also be run with buffered aqueous solvents without difficulty, since the solvent background is referenced against all the sample collection data, and the buffer spectrum would be ratioed out. The only problem that might arise relates to gradient elution with buffered solvents. In such a case the ratioing process would be more complicated and might not be sufficiently useful to allow minor components to be identified unambiguously.

RESULTS AND DISCUSSION

The above-described interface can accommodate aqueous solvent systems with up to 100% water, although 80% water is a practical limit for the chromatography. Normal phase (and nonaqueous reversed-phase) chromatography can be carried out with the system as well by simply switching off the flow of the post-column reactant and, of course, using an appropriate column. The splitter interface for analytical columns and the microbore interface have approximately the same flow rates directed toward the postcolumn reactor and diffuse reflectance cell. The operating conditions for the postcolumn region will differ only slightly, and the selection of analytical or microbore HPLC is as easy as connecting the appropriate tubing.

Caffeine has been found to be a good test sample for this aqueous reversed-phase system, as it has a very distinct, rather intense infrared spectrum, and it elutes in reasonably short times in a variety of solvent systems. Shown in Figure 3 is an elution of caffeine from an analytical reversed-phase column using 20% water and 80% methanol. The "stacked plot" is a real-time display of the infrared data showing the peak elution. The complete infrared spectrum on the right was obtained from data file 312 on the left and is displayed to emphasize that there are no absorbances due to water. Practical detection limits for caffeine are approximately 250 ng of sample directed toward the DRIFT accessory, but, typically, samples of the order of a microgram are required to produce high quality infrared data.

The diffuse reflectance infrared spectrum of 0.5 µL of water is shown in Figure 4. It is evident from the figure that a trace of water would blank out of the majority of the infrared data for a solute. Inspection of the spectrum of caffeine in the previous figure shows that the postcolumn reaction has eliminated all of the water in the mobile phase and that the other solvents were removed in the final drying stage of the interface.

The real-time infrared stacked plot of the separation of niacin and nicotinamide from a 60% water/40% methanol solvent system on a microbore column is shown in Figure 5 [25]. Both peaks of the liquid chromatograph can be identified from the infrared spectra as they "grow" in and out of the plots. The niacin covers infrared data files 328–331, while nicotinamide is found in data files 331–334. It is clear that there is some overlap of the two components in file 331, but the infrared data collected from that run can be used to identify both components. Figure 6 shows that the spectrum of nicotinamide agrees very well with the reference data even out to 4000 cm^{-1}, beyond the frequency where any residual water would give interfering peaks. This study, however, points out an interesting phenomenon insofar as the infrared spectra of niacin and many other solids can exhibit differences in relative intensities of certain bands depending upon the particular crystalline state of the material.

Another example of the data that can be obtained from the interface is shown in Figure 7, where the three components of an analgesic preparation were separated in a 50% water/50% methanol solvent on a microbore column. The reconstructed infrared chromatogram is shown on the right of the stacked plot. All three peaks are evident, and the abscissa (which is a function of time) indicates which data files contain particular eluate spectra.

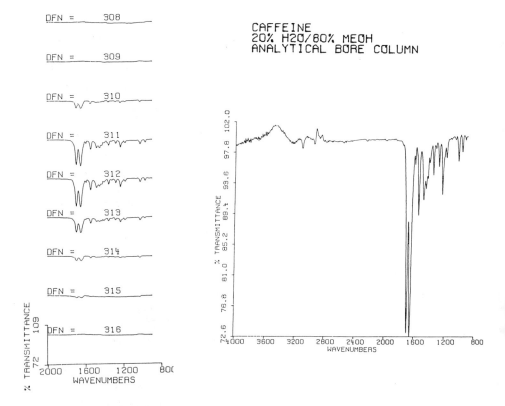

Figure 3 Partial stacked plot of 7.5 µg of caffeine eluting from the splitter in water/methanol (20/80) and the caffeine spectrum taken from data file 312. Used with permission, Ref. 26.

Figure 8 shows the separation of two steroids (cortisone and corticosterone) from a 40% water/60% methanol solvent system on an analytical column [26]. The reconstructed infrared chromatogram is shown on the right. Again both peaks are strongly evident, and Figure 9 shows the comparison of the HPLC/FT-IR data with infrared reference spectra. Again a very good match is observed.

Shown in Figure 10 is the separation of a three-component anticonvulsant from a 30% water/70% methanol solvent system on a microbore column. Figure 11 shows the data collected from this run, but the identification has not been made because suitable reference spectra were unavailable.

Figure 12 shows the separation of acetaminophen and aspirin from a 50% water/50% methanol solvent on an analytical column. Shown in Figure 13 is a comparison of the reconstructed infrared chromatogram and the ultraviolet trace (254 nm) from a conventional HPLC detector positioned before the postcolumn reactant. Excellent agreement was obtained, and bandwidths are comparable in the two chromatograms.

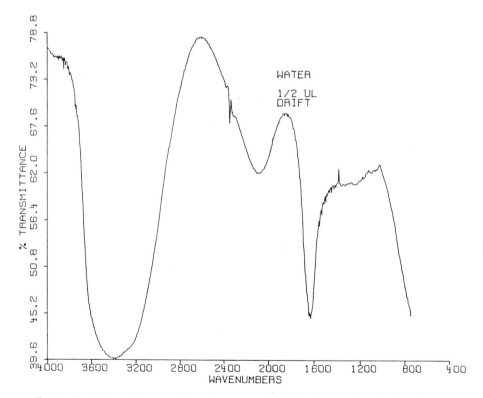

Figure 4 Diffuse reflectance infrared spectrum of 0.5 µL of water deposited on a train.

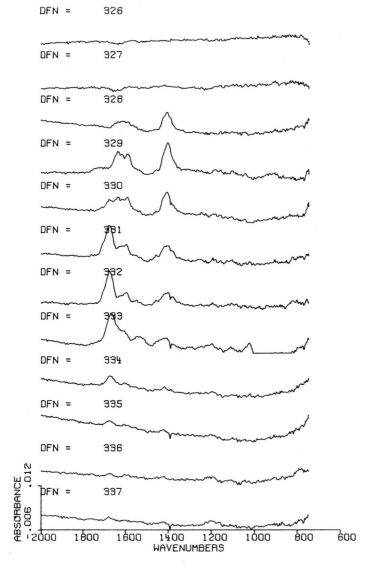

Figure 5 Partial stacked plot of 2.5 µg of niacin and 2.5 µg of nicotinamide eluting in water/methanol (60/40) and deposited on a train. Column flow rate was 50 µL/min. Used with permission, Ref. 25.

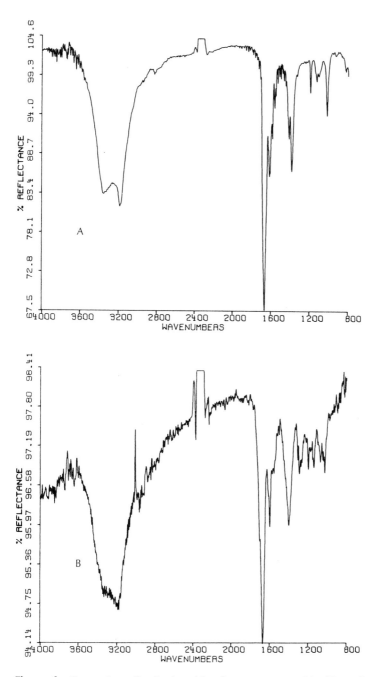

Figure 6 Comparison of a nicotinamide reference spectrum (a) with an eluate spectrum (b) taken from one compartment of the train (4000–800 cm⁻¹). Used with permission, Ref. 25.

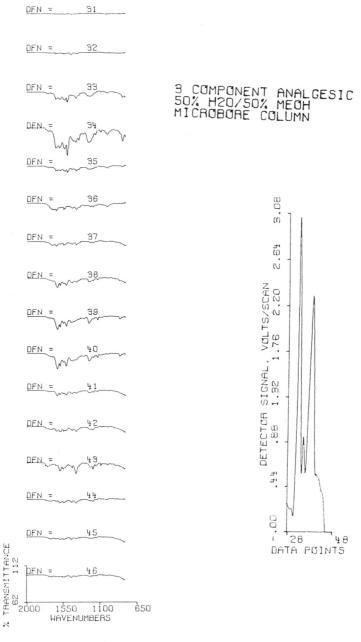

Figure 7 Partial stacked plot and infrared chromatogram for a three-component analgesic preparation eluting in water/methanol (50/50), and an infrared chromatogram showing three peaks. Column flow rate was 50 µL/min; DMP flow rate was 150 µL/min.

Design and Application of HPLC/FT-IR

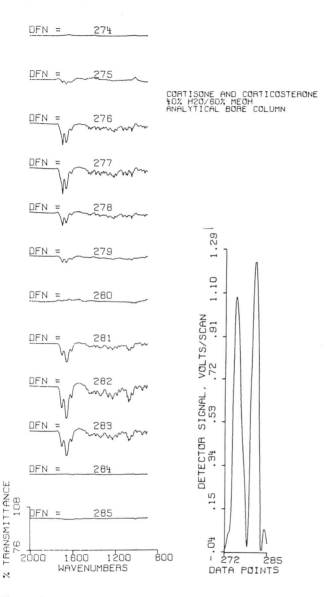

Figure 8 Partial stacked plot of 17 μg of cortisone and 17 μg of corticosterone eluting from the splitter in water/methanol (40/60), and in infrared chromatogram showing two peaks. Column flow rate was 1.0 mL/min, and the DMP flow rate was 0.13 mL/min. Used with permission, Ref. 26.

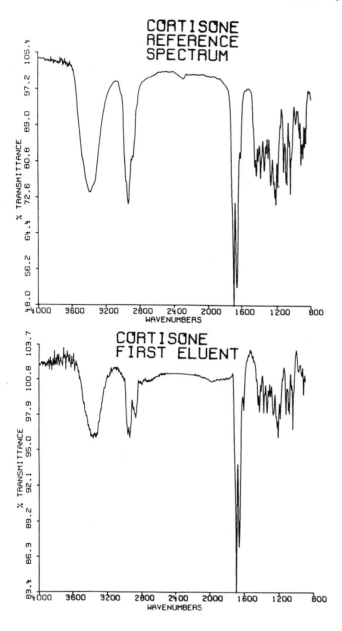

Figure 9 Comparison of cortisone and corticosterone reference spectra with eluate spectra taken from the train. Used with permission, Ref. 26.

Design and Application of HPLC/FT-IR

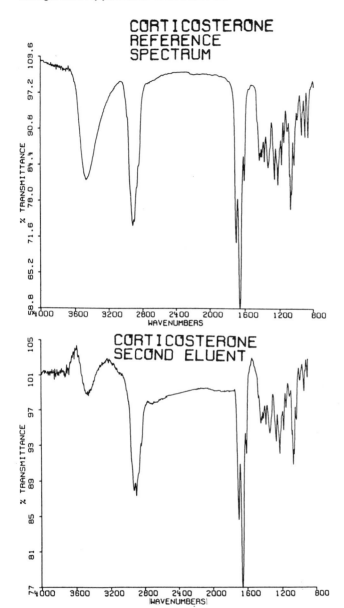

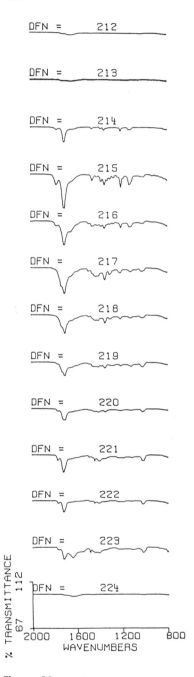

Figure 10 Partial stacked plot of a three-component anticonvulsant eluting in water/methanol (30/70). Column flow rate was 50 µL/min; DMP flow was 90 µL/min.

Design and Application of HPLC/FT-IR

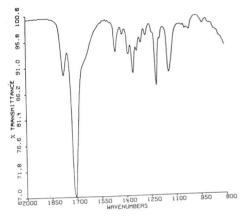
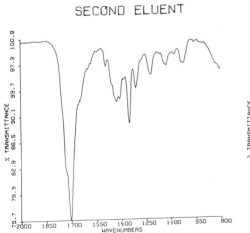
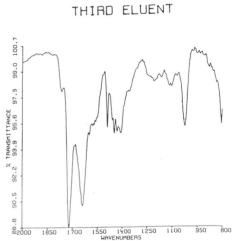

Figure 11 Infrared spectra of the three components of the anticonvulsant taken from Figure 10.

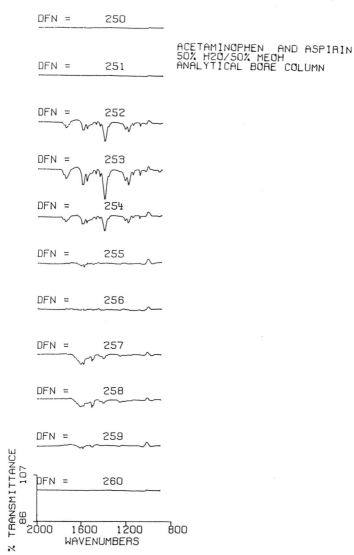

Figure 12 Partial stacked plot of acetaminophen and aspirin eluting from the splitter in water/methanol (50/50).

Design and Application of HPLC/FT-IR

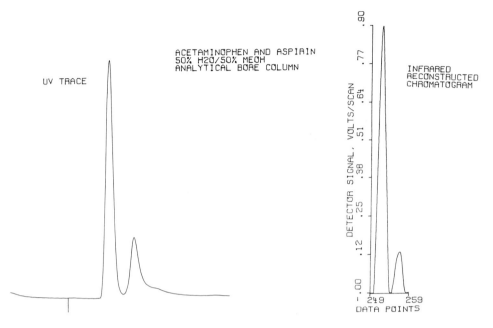

Figure 13 Ultraviolet (254 nm) and infrared chromatograms of acetaminophen and aspirin taken from Figure 12.

The interface described above is quite versatile, as previously mentioned. It can be used with normal- or reversed-phase HPLC and with analytical or microbore columns. As the examples indicate, for aqueous reversed-phase chromatography, changes in the percentage composition of the solvent system can be accommodated in the interface by changing the DMP flow, and this is particularly easy with a microprocessor-based controller. Such is the method employed for gradient elution HPLC/FT-IR. A complete description of this work can be found in Ref. 40. Shown therein is an application of reversed-phase microbore HPLC/FT-IR for the pharmaceutical steroids corticosterone, hydrocortisone, progesterone, and testosterone in which a linear gradient from 35% water to 100% methanol was used [40].

The interface is compact, and the sample trains are relatively easy to handle. In this regard, the ability to operate the interface on line or off line provides flexibility, in that it is not necessary permanently (physically) to couple the HPLC and infrared equipment. Although all the examples shown above involved water/methanol solvent systems, other solvent systems work as well. The only important characteristic of a solvent (other than water) is that it must be volatile. The postcolumn reactant (2,2- dimethoxypropane) is specific to water, and the reaction is very rapid.

As with any analytical technique, there are trade-offs in HPLC/FT-IR. The deposition time per compartment of the train is related to the chromatographic resolution, and a shorter deposition time (which gives better chromatographic resolution) results in a smaller amount of sample per compartment that would be available for infrared analysis. So the infrared sensitivity depends on the desired chromatographic resolution. Another factor affecting the infrared sensitivity relates to the method and efficiency of the deposition process. The *absolute* detection limits for a sample like caffeine in the diffuse reflectance accessory and infrared spectrometer used in our experiments are of the order of 20 ng when the sample is carefully deposited (manually) as a 2-mm-diameter spot. With the HPLC interface and its deposition method, the spot size is larger, and at least *half* the deposited sample is not detected by the infrared spectrometer. And while this still enables adequate detection of submicrogram quantities (injected in the microbore system), there is certainly room for improvement in the interface.

Current efforts in our laboratory and others are aimed at improving the sensitivity of the infrared detection scheme and improving the deposition and solvent elimination procedures. With improved detection limits, applications in detecting minor impurities in pharmaceuticals will become more attractive. In the meantime, however, it is possible to utilize infrared detection for identifying nanogram quantities of HPLC eluates. For completeness, it is also worthwhile to mention that supercritical fluid chromatography/infrared (SFC/IR) interfaces that can be used with either flow cells or solvent-elimination detection have been constructed [41-43], but because of solvent selection it remains to be seen whether SFC can be as widely applicable to pharmaceutical problems as HPLC already is. In a recent article, however, Griffiths and coworkers have suggested that a single solvent-elimination interface might be used for GC/IR, HPLC/IR, and SFC/IR [44]. The interface utilizes microscope optics to analyze very small sample "spots" that are immobilized on a cooled ZnSe plate. This interface offers many interesting possibilities.

We have chosen but a few examples of the HPLC/FT-IR data that can be obtained with the interface developed in our laboratories, and it appears to be appropriate to pharmaceutical formulations. Our access to controlled substances that might be found in a pharmaceutical laboratory was limited, but we believe that additional applications could be developed to complement the widespread use of HPLC alone.

ACKNOWLEDGMENTS

The authors are grateful to Dr. James A. Smith, Dr. Keith G. Whitehead, and Dr. Roger C. Kenton for their efforts in performing the experimental work discussed above.

REFERENCES

1. D. W. Vidrine, Liquid chromatography detection using FT-IR, *Fourier Transform Infrared Spectroscopy*, Vol. 2 (J. R. Ferraro and L. J. Basile eds.), Academic Press, New York, p. 129 (1979).
2. D. W. Vidrine and D. R. Mattson, A practical real-time Fourier transform infrared detector for liquid chromatography, *Appl. Spectrosc.*, *32*: 502 (1978).

3. D. W. Vidrine, Use of subtraction techniques in interpreting on-line FT-IR spectra of HPLC column eluates, *J. Chrom. Sci.*, *17*: 477 (1979).
4. K. H. Shafer, S. V. Lucas, and R. J. Jakobsen, Application of the combined techniques of HPLC/FT-IR, GC/FT-IR, and GC/MS to the analysis of real samples, *J. Chrom. Sci.*, *17*: 464 (1979).
5. R. S. Brown, D. W. Hausler, L. T. Taylor, and R. O. Carter, Fourier transform infrared spectrometric detection in size-exclusion chromatographic separation of polar synfuel material, *Anal. Chem.*, *53*: 197 (1981).
6. R. S. Brown and L. T. Taylor, Microbore liquid chromatography with flow cell Fourier transform spectrometric detection, *Anal. Chem.*, *55*: 1492 (1983).
7. K. Jinno, C. Fujimoto, and G. Uematsu, Micro-HPLC/FTIR, *Am. Lab. (Fairfield, Conn.)*, *16*(2): 39 (1984).
8. C. C. Johnson and L. T. Taylor, Zero dead volume flow cell for microbore liquid chromatography with Fourier transform infrared spectrometric detection, *Anal. Chem.*, *56*: 2642 (1985).
9. C. C. Johnson, J. W. Hellgeth, and L. T. Taylor, Reversed-phase liquid chromatography with Fourier transform infrared spectrometric detection using a flow cell interface, *Anal. Chem.*, *57*: 610 (1985).
10. P. A. Wilks, Jr. Sampling method makes on-stream IR analysis work, *Ind. Res. Dev.*, *24*(9): 132 (1983).
11. M. Sabo, J. Gross, J. Wang, and I. E. Rosenberg, On-line high performance liquid chromatography/Fourier transform infrared spectrometry with normal and reverse phase using an attenuated total reflectance flow cell, *Anal. Chem.*, *57*: 1822 (1985).
12. P. R. Griffiths and C. M. Conroy, Solvent elimination techniques for HPLC/FT-IR, *Adv. Chromatogr.*, *25*: 105 (1986).
13. D. Kuehl and P. R. Griffiths, Novel approaches to interfacing a high performance liquid chromatograph with a Fourier transform infrared spectrometer, *J. Chrom. Sci.*, *17*: 471 (1979).
14. D. T. Kuehl and P. R. Griffiths, Microcomputer-controlled interface between a high performance liquid chromatograph and a diffuse reflectance infrared Fourier transform spectrometer, *Anal. Chem.*, *52*: 1394 (1980).
15. K. Jinno, C. Fujimoto, and Y. Hirata, An interface for the combination of micro high-performance liquid chromatography and infrared spectrometry, *Appl. Spectrosc.*, *36*: 67 (1982).
16. C. Fujimoto, K. Jinno, and Y. Hirata, Liquid chromatography-spectrometry with the buffer-memory technique, *J. Chromatogr.*, *258*: 81 (1983).
17. C. M. Conroy, P. R. Griffiths, and K. Jinno, Interface of a microbore high-performance liquid chromatograph with a diffuse reflectance Fourier transform infrared spectrometer, *Anal. Chem.*, *57*: 822 (1983).
18. J. J. Gagel and K. Biemann, Continuous recording of reflection-absorption Fourier transform infrared spectra of the effluent of a microbore liquid chromatograph, *Anal. Chem.*, *58*: 2184 (1986).
19. C. M. Conroy, P. R. Griffiths, P. J. Duff, and L. V. Azarraga, Interface of a reversed-phase high-performance liquid chromatograph with a diffuse reflectance Fourier transform infrared spectrometer, *Anal. Chem.*, *56*: 2636 (1984).
20. J. M. Brackett, L. V. Azarraga, M. A. Castles, and L. B. Rogers, Matrix materials for diffuse reflectance Fourier transform infrared spectrometry of substances in polar solvents, *Anal. Chem.*, *56*: 2007 (1984).

21. M. A. Castles, L. V. Azarraga, and L. A. Carreira, Continuous, on-line interface for reversed-phase microbore high performance liquid chromatography/diffuse reflectance infrared Fourier transform analysis, *Appl. Spectrosc.*, *40*: 673 (1986).
22. C. Fujimoto, T. Oosuka, and K. Jinno, A new sampling technique for reversed-phase liquid chromatography/Fourier-transform infrared spectrometry, *Anal. Chim. Acta*, *178*: 159 (1985).
23. J. J. Gagel and K. Biemann, Continuous infrared spectroscopic analysis of isocratic and gradient elution reversed-phase liquid chromatography separations, *Anal. Chem.*, *59*: 1266 (1987).
24. R. M. Robertson, J. A. deHaseth, J. D. Kirk, and R. F. Browner, MAGIC- LC/FT-IR spectrometry: Preliminary Studies, *Appl. Spectrosc.*, *42*: 1365 (1988).
25. K. S. Kalasinsky, J. A. S. Smith, and V. F. Kalasinsky, Microbore high-performance liquid chromatography/Fourier transform infrared interface for normal- or reversed-phase liquid chromatography, *Anal. Chem.*, *57*: 1969 (1985).
26. V. F. Kalasinsky, K. G. Whitehead, R. C. Kenton, J. A. S. Smith, and K. S. Kalasinsky, HPLC/FT-IR interface for normal- or reversed-phase analytical columns, *J. Chrom. Sci.*, *24*: 273 (1987).
27. H. C. Dorn, Proton NMR: a new detector, *Anal. Chem.*, *56*: 747A (1984).
28. D. A. Laude, Jr., R. W. K. Lee, and C. L. Wilkins, Reversed-phase high-performance liquid chromatography/nuclear magnetic resonance spectrometry separations of biomolecules with 1-1 hard pulse solvent suppressions, *Anal. Chem.*, *57*: 1464 (1985).
29. J. F. Holland, C. G. Enke, J. Allison, J. T. Stults, J. D. Pinkston, B. Newcome, and J. T. Watson, Mass spectrometry of the chromatographic time scale: realistic expectations, *Anal. Chem.*, *57*: 997A (1983).
30. P. R. Griffiths, J. A. deHaseth, and L. V. Azarraga, Capillary GC/FT-IR, *Anal. Chem.*, *55*: 1361A (1983).
31. G. T. Reedy, D. G. Ettinger, J. F. Schneider, and S. Bourne, High-resolution gas chromatography/matrix isolation infrared spectrometry, *Anal. Chem.*, *57*: 1602 (1985).
32. C. P. Tsai, AZ. Sahil, J. M. McGuire, B. L. Karger, and P. Vouros, High-performance liquid chromatography/mass spectrometric determination of volatile carboxylic acid using ion-pair extraction and thermally induced alkylation, *Anal. Chem.*, *58*: 2 (1986).
33. T. R. Covey, J. B. Crowther, E. A. Dewey, and J. D. Henion, Thermospray liquid chromatography/mass spectrometry determination of drugs and their metabolites in biological fluids, *Anal Chem.*, *57*: 474 (1985).
34. J. B. Crowther, T. R. Covey, E. A. Dewey, and J. D. Henion, Liquid chromatographic/mass spectrometric determination of optically active drugs, *Anal. Chem.*, *56*: 2921 (1984).
35. J. G. Stroh, J. C. Cook, R. M. Milberg, L. Brayton, T. Kihara, Z. Huang, and K. L. Rinehart, Jr., On-line liquid chromatography/fast atom bombardment mass spectrometry, *Anal. Chem.*, *57*: 985 (1985).
36. K. H. Shafer, P. R. Griffiths, and S. -Q. Wang, Sample transfer accessory for thin-layer chromatography/Fourier transform infrared spectrometry, *Anal. Chem.*, *58*: 2708 (1986).
37. R. A. Bredeweg, L. D. Rothman, and C. D. Pfeiffer, Chemical reactivation of silica columns, *Anal. Chem.*, *51*: 2061 (1979).
38. D. S. Erley, 2,2-dimethoxypropane as a drying agent for preparation of infrared samples, *Anal. Chem.*, *29*: 1564 (1979).

39. V. F. Kalasinsky, J. A. S. Smith, and K. S. Kalasinsky, Rapid and convenient sampling accessory for diffuse reflectance spectroscopy. *Appl. Spectrosc.*, *39*: 552 (1985).
40. R. C. Kenton, Developments and applications of microbore HPLC/FT-IR and vibrational studies of some substituted three-membered rings and organosilanes, Ph. D. dissertation, Mississippi State University (1988).
41. K. H. Shafer, S. L. Pentoney, Jr., and P. R. Griffiths, Superficial fluid chromatography/Fourier transform infrared spectrometry with an automatic diffuse reflectance interface, *Anal. Chem.*, *58*: 58 (1986).
42. S. B. French and M. Novotny, Xenon, a unique mobile phase for supercritical fluid chromatography, *Anal. Chem.*, *58*: 164 (1986).
43. G. L. Pariente, S. L. Pentoney, Jr., P. R. Griffiths, and K. H. Shafer, Computer-controlled pneumatic amplifier pump for supercritical fluid chromatography and extractions, *Anal. Chem.*, *59*: 808 (1987).
44. A. M. Haefner, K. L. Norton, P. R. Griffiths, S. Bourne, and R. Curbelo, Interfaced gas chromatography and Fourier transform infrared transmission spectrometry by eluite trapping at 77K, *Anal. Chem.*, *69*: 2441 (1988).

Part Three
Automation in Pharmaceutical Analysis

8
Application of HPLC to Dissolution Testing of Solid Dosage Forms

William A. Hanson

Hanson Research Corporation, Chatsworth, California

UNIT OPERATIONS IN DISSOLUTION TESTING

One can more clearly appreciate the use of HPLC in a dissolution test protocol when the test is separated into distinct unit operations. A clear understanding of unit operations in the dissolution test enables one to proceed with automation in an organized pattern. One may concentrate on the operation that is more time-consuming or likely to provide the best cost/accuracy trade off [1]. The author suggests the following as a guideline for unit operations in dissolution testing (see Figure 1).

1. *Dissolution.* Set up, start, provide compendia compliance with dissolving the dosage form under defined conditions.
2. *Sampling.* Withdraw sample aliquots from the proper position in the dissolution vessel at the exact time specified in the dissolution test protocol.
3. *Sample preparation.* Condition, archive (store), modify (column, dilute, add reagent, etc.).
4. *Analysis of Sample.* Determine concentration of target ingredient(s) as a function of time.
5. *Data Reduction.* Record key values, manipulate to correct input errors, determine statistical information, plot graphs. This function may also include computer-directed control of the system operation.

Since this chapter is devoted to the use of HPLC in dissolution testing, unit operations 3 (sample preparation) and 4 (sample analysis) will be emphasized.

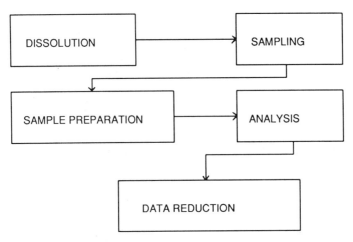

Figure 1 Author's concept of dissolution unit operations in diagram form.

WHY HPLC IN DISSOLUTION TESTING

It is convenient to separate sample preparation and detection into two unit operations. In practice, this distinction may be significant. Some analytical methods may not require sample preparation. An obvious example is UV/VIZ spectrophotometric determination of absorbance. Generally no sample modification other than filtering and possible dilution is necessary. Direct reading without sample preparation is even used for multicomponent analysis using scanning or diode array technologies.

Other methods of analytical detection may or may not require sample modification for successful application. Sample preparation is inherent in HPLC with a choice of detector. It is convenient to consider HPLC as a combination of sample preparation with a suitable detector.

Sample preparation may be desirable for any number of reasons; (a) there may be interference by excipients or impurities that have no analytical interest, (b) active ingredients may overlap and interfere in the detection system, (c) the analytical method may not detect an unmodified molecule, or (d) the target ingredient may have too low a concentration for the detection method chosen.

Unmodified UV absorption is still the most universally used dissolution analytical procedure. It is simple, direct, and not too expensive. Extensive dissolution programs for data reduction are available. It may be particularly attractive where a single component determination after a short dissolution period is required. Aspirin is an example of such an application.

However, with the growing tendency towards controlled release dosage forms with complex release mechanisms, sample preparation requirements are likely to increase until they occupy the major position in dissolution text protocols.

DISSOLUTION TESTING PROBLEMS SOLVED WITH HPLC

Sample preparation may be used to handle a variety of problems in complex dissolution systems. Some of these are listed and addressed in the following paragraphs.

Low Concentrations of Active Ingredients

One is constrained from selecting a convenient sample concentration because the potency of the dosage form (and perhaps the dissolution volume in a specified method) is already established.

Monitoring Excipients and Degradation Products

The tendency to employ exotic excipients, such as various polymeric substances in controlled release formulations, is increasing. The development of interference and degradation products during the dissolution test of these items is a serious problem.

Both of these factors may be uniquely solved by using HPLC as a sample preparation procedure. An early example of the use of HPLC to handle both problems is demonstrated in the dissolution test of fludrocortisone acetate tablets in dosage units of 100 µg. The 100 µg dosage established a predetermined maximum concentration that could not be altered except by reducing the dissolution media volume. In this example, the volume was reduced by a modification of Method 1 until a measurable concentration of the dissolved drug was possible [2]. Even with this modification, the concentrations were in the range of 1 µg/ml, and the sensitivity of alternative methods was inadequate. HPLC provided a sensitivity of 0.5 µg/ml, however, and was selected.

Further, fludrocortisone acetate I degrades to fludrocortisone acetate II-IV. These degradation products were easily monitored by HPLC separation [3]. Thus in this example both problems, below-detectable concentration and appearance of degradation products, were solved by selecting HPLC as the analytical method.

The increased use of excipients and polymers in the formulation of controlled release drugs multiplies the probability of interference products. This probability increases still more when patch delivery systems are designed. Monitoring the presence of such molecules during the dissolution test is important. The possibility of such products developing during the dissolution test period is significant when such tests last over extended time periods of hours or days. Since dissolution data are based on the total amount of active ingredient released into solution during the total test time, degradation products may arise that significantly alter data. Their identification therefore is essential to accurate dissolution analysis. HPLC is admirably suited to this task, since it can simultaneously quantify both the parent component and the degradation products.

Multicomponent Analysis

Multicomponent dosage forms are conveniently and accurately analyzed when subjected to the sample preparation operation known as HPLC. Proper column selection enables the analyst to separate ingredients adequately with minute differences or broad interfering characteristics. These interferences may be very troublesome without sample preparation.

Such analysis is commonly performed by scanning UV spectrophotometers or programmed instruments that read more than one wavelength. When peaks of target compounds are sufficiently separated to minimize interference this is generally satisfactory. Dissolution samples may be checked and individual components recognized and measured. The introduction of the diode array speeds up, expands, and provides statistical data enabling accurate analysis of several components. Since the introduction of the diode array, the use of UV absorbance for multicomponent analysis in dissolution testing has become more popular.

The usefulness of the diode array for multicomponent analysis, however, is limited to selected components whose absorbance is measured over a series of data points and a least square fit that makes it possible to analyze up to six or seven components [4].

It is significant to point out that commercially available hardware and software for these techniques applied to dissolution are as of this writing rare. The most popular multicomponent dissolution software is offered by a manufacturer [4] whose primary business is computers. One may ponder the ultimate cost effectiveness of substituting software intensive procedures for the comparatively simple LC separation used as sample preparation prior to detection.

Broad Choice of Detection Methods

Although a majority of drugs can be identified through UV absorbance there are some that are more conveniently and accurately assayed by other procedures. These may include a wide variety of methods such as pH, specific ion, and polarimetry. But any of these may be more satisfactory when preceded by LC sample preparation. UV is the most common, but not the only, detector used with HPLC.

Small Sample Volumes

Since it concentrates as well as separates, HPLC may be used in dissolution testing where extremely small sample aliquots are specified, e.g., in the µl order of magnitude. The use of such small volumes is inclined to increase in the next decade as dissolution testing finds greater application in the field of percutaneous absorption studies.

Transdermal Dosage Dissolution Testing

This area of dissolution testing includes patch delivery systems as well as ointments. Both of these are likely to become subject to more detailed regulatory standards in the future.

The obvious advantage of patch delivery dosage for drugs requiring constant blood levels dictates the gradual replacement of oral delivery in the future. Even sustained release dosage forms are real-time limited to their duration in the alimen-

Application of HPLC to Dissolution Testing

tary tract, e.g., 4–12 hr. Patch delivery, on the other hand, may be theoretically effective for days, weeks, and even months.

These criteria pose severe demands on dissolution testing as a quality control procedure. A 10 mg daily oral dose, for example, translates into a release rate of less than 0.5 mg/hr.

Dissolution testing of patch delivery systems involves transfer through both the membrane enclosing the patch and the skin or other natural barrier. Two distinct methods of dissolution (unit operation #1 in this chapter) testing are used. While many suggested techniques are being studied for in vitro patch dissolution, a recent supplement to the USP/NF describes three that have been subject to collaborative studies [5]. The rate of transfer through the skin, however, involves studies of percutaneous absorption with special glass cells using synthetic membranes or skin [6, 7].

Samples from any of these methods are best analyzed by using the sample preparation separations of HPLC. Each one either requires extremely small volume (microliter range) samples and/or exhibits extremely low concentrations that may not be handled by other methods because of sensitivity limitation.

It should also be mentioned that transdermal dosage forms, whether patch or ointment, will by nature have two other characteristics: (a) excipients and polymers designed to enable the dose to diffuse through the membrane and/or skin at a controlled rate and (b) a requirement to maintain chemical stability during dissolution testing over days, weeks, and even months as compared to hours for oral dosage forms. Each of these causes problems in dissolution testing analysis: (a) the monitoring of the amount and stability of multiple excipients and (b) the presence of degradation products that interfere with the detection of the active drug as a result of prolonged dissolution test time.

The evidence suggests that the sample preparation inherent in HPLC is the method of choice for analysis in the dissolution testing of ointments, patches, and transdermal dosage forms.

Summary of Reasons for Selecting HPLC

1. Concentration and separation makes possible analysis of amounts undetectable by other non-sample-preparation methods.
2. HPLC enables routine monitoring of some key excipients, which are increasingly necessary to consider as the popularity of modified release dosage forms grows.
3. It allows measurement and monitoring of known and unknown degradation products that may develop in the extended time phases of modified release preparation dissolution testing.
4. It is a cost effective and accurate system to handle the analysis of dissolution samples containing several target components.
5. It can prepare samples for a wide variety of detectors.

6. Its procedures require far less volume of sample than other methods.
7. Its inherent characteristics lend themselves much better than other methods for the dissolution testing of ointments and transdermal patch dosage forms.

The author concludes that HPLC will grow in the next decade as the analytical method preferred in a substantial majority of dissolution testing protocols.

HPLC HARDWARE AND PROCEDURES IN DISSOLUTION TESTING

Success in the application of LC to pharmaceutical analysis depends on the selection of the proper column, mobile phases, sample condition, time and other parameters that are adequately covered elsewhere in this volume. These methods and techniques are not the subject of this chapter.

In reviewing the application of HPLC to dissolution test protocols we assume that these matters have been determined. We have pointed out in this chapter the present and probable future advantages of HPLC sample preparation and detection in dissolution test procedures.

A brief review of the hardware and procedures currently used in conjunction with HPLC in dissolution testing follows. An attempt will be made to limit our discussion to the practical aspects of selecting and operating these systems. Certain precautions and hints to avoid extra expense and difficulties are covered.

If readers will glance again at the unit operations in dissolution testing listed and diagrammed at the start of this chapter, they will realize that as a system expands each unit requires interfacing. It is certainly wise to select modular components rather than one cabinet containing everything. This allows for gradual expansion as needs grow and also provides protection against catastrophic system breakdown. In critical flow situations when production lines depend on uninterrupted dissolution testing, spare modules may be inserted in a few minutes by factory personnel without waiting for instrument service to arrive.

This factor is of increasing importance as dissolution testing for QC becomes, in effect, a part of the production line. Most HPLC suppliers recognize this problem and offer modular components for their systems.

Interfacing the various unit operations in dissolution testing, however, is often left to the improvisation of the laboratory technician. It is exasperating to install an expensive analytical unit only to find that one must improvise the necessary interfaces.

Sampling

Withdrawing sample aliquots from the dissolution flask is the second unit operation. In the simplest case this is accomplished by hand, using syringes with cannulae to locate manually the proper sampling position in the dissolution vessel. Precautions are necessary with this technique to avoid misplacement of sample point, errors in

Application of HPLC to Dissolution Testing

timing of sample withdrawal, and errors resulting from ejecting the sample through a filter [8]. These samples are temporarily stored in vials for later injection into an HPLC system. Evaporation from these small volumes can introduce significant error.

Automated sampling is more accurate than hand methods. It can precisely locate the sample point, exactly repeat sampling times, and accurately measure the volume withdrawn. Any commercial automated sampling apparatus should supply tubing, filters, and holders that make it unnecessary for the technician to improvise with tape-held probes or other methods. This tubing should easily locate the proper sampling position and be of a size and shape to avoid disturbance of dissolution flow patterns [9]. Alternatively it may be automatically removed after the sampling sequence. Automatic removal can be achieved by robots [10] or by sample probes that automatically are inserted only during the sampling period [11].

Filter Selection

Sorption on tubing or filters is often encountered, and it should be checked by simple tests prior to selection and installation. Special test procedures are suggested [8] to avoid errors from sorption of dissolution sample ingredients on filters. Similar tests can be applied to tubing. When filters are reused (as in successive samplings) they should be tested by successive sampling through the same filter using 100% standard and blank solutions. Recovery of the standard should be within 95-101%, and no more than 5% should appear in the blank. Greater tolerances indicate possible cumulative sorption. One solution is to saturate the filters and lines before use. A wide range of filter materials is available for dissolution testing ranging in composition through cellulosic materials, polycarbonate, sintered polyethylene, polyfluorocarbon, stainless steel, and ordinary glass wool.

Those using HPLC methods with these automated sampling devices should recognize that automated samplers generally operate at low pressure differentials (15 psi or less) and are not capable of providing satisfactory flow rates with some filters with less than 1 micron pore size. If smaller pore sizes are required, the analysts are advised to consider the use of accessory filters in the high pressure sample injection portion of their HPLC.

Particulates may accumulate on the surface of filters. Aside from clogging, these particulates are subjected to a disturbingly higher flow shear rate, leading to errors in dissolution data. These errors may be significant and not easily detected, since they may also occur in manual sampling [8].

This phenomenon was first brought to the author's attention at a Pan American Health Organization seminar in dissolution in Panama City in 1985. The concentration of individual dissolution samples exceeded 100% of its theoretical possibility, but it dropped to nominal values when the particulates finally dissolved [12]. Both this error and clogging may be avoided by a reverse filter purge before each sample cycle. This function should be a first consideration for a person selecting sampling filter procedures in dissolution.

Tubing Material

Many drugs exhibit sorption on plastic tubing. One of the first observations of this phenomenon in analytical trains was observed in testing digoxin. That problem was solved by using a modification of the dissolution sampling process, introducing solvents into the sample stream [13]. Such chemical solutions suggest further an emphasis on the use of HPLC, which easily monitors the presence and action of these added reagents.

The use of peristaltic pumps in dissolution sampling requires the insertion of flexible plastic tubing with a "memory." At the time of this writing, polyfluorocarbon tubing such as Teflon exhibits the least sorption of these sensitive drugs, but this material has not been successfully used in peristaltic pumps.

Alternative fluid movement methods are available, however. Pumps with all wetted parts from Teflon and stainless steel are commercially available [14]. Application of vacuum/pressure sequences under microprocessor control has been successfully employed for several years in dissolution sampling equipment [15]. Polyfluorocarbon and stainless wetted valves are used in these systems.

Surfactants

There is an increasing tendency to use surfactants such as sodium lauryl sulfate compounds in controlled release dosage forms. The change in surface tension of the sample line may introduce severe problems in fluid transfer systems. Bubbles form in flow cells or the dissolution flasks become "bubble machines." This problem is solved with unique sample moving and dispensing systems that avoid the introduction of sudden changes of pressure differentials or atmospheric contamination in the sample train [11].

Sample Injection

A precise volume of sample must be injected under high pressure into modern HPLC columns. The standard procedure is to fill a measured loop in an injector valve with precise volume and without carryover. Injector valves, manual and pneumatic or electrically actuated, are available from several manufacturers. The fluid dynamics of filling the loop are highly complicated and subject to errors that are discussed in the technical bulletins of one manufacturer [16]. Some of these problems are avoided with a unique patented method of measuring and injecting the sample [17].

Automating Sample Injection

Several attempts have been made to provide dissolution samples for a sample loop valve directly as the dissolution test proceeds. In cases known to the author, all such attempts have been discarded. The technology is simple, being very similar to delivering a dissolution sample to a UV flow cell. However, the assay time required for each sample in the HPLC column seldom synchronizes with the time specified in the sampling intervals for the dissolution test.

Application of HPLC to Dissolution Testing

Therefore as of this writing successful interface of dissolution sampling unit operation involves storage of samples for subsequent HPLC assay. The samples are delivered to a sample collector and subsequently fed into an HPLC system.

Asynchronicity is not the only reason for this practice. Generally, expensive HPLC equipment should not stand idle while dissolution sampling proceeds sometimes only at hourly intervals. Indeed, some laboratories are so organized that HPLC analysis is typically assigned to a totally different task group from dissolution testing, often is separate facilities.

This has led to the practice of storing the dissolution samples in transferable carousels such as the WISP TM autoinjector carousel [17]. Dissolution sampling systems are available to fill automatically these and other carousels used by manufacturers of automated HPLC injection equipment [11].

Development of complete systems that allow dissolution samples to be directly deposited into automated injection systems at the required time intervals for dissolution sampling, but using the real time between sample collection to autoinject, have also been developed and are commercially available [18]. Such a system (see Figure 2) allows completely unattended dissolution/HPLC analysis to proceed for all types of dissolution testing. It is computer directed, and the output data lags the dissolution test sequence only by the real time necessary to accommodate HPLC assay time vs. sampling time. This system is also available for transdermal Franz cell type (7) studies.

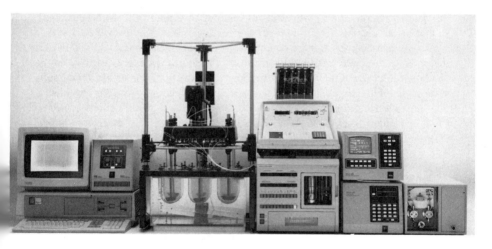

Figure 2 Fully automated HPLC dissolution system. Samples from the test stand are automatically injected into sealed vials of the autoinjector and programmed to feed through the column and detector. This system is computer directed and capable of processing up to 96 samples per run with variable sample times and includes print out of dissolution data. (Courtesy of Waters Associates, Milford, Mass.)

DATA REDUCTION AND COMPUTER CONTROL

The world of computer technology seems to offer a solution to all problems. The advent of mainframe data storage and control in major pharmaceutical companies means that analytical data must interface with such facilities. On a practical plane, however, one should separate the functions of instrument process control from data manipulation and storage requirements.

Operating commands, such as those to turn a pump on, determine how many seconds are devoted to sample withdrawal, or to index a collector position, may be highly specific, and they may vary with each dissolution test depending upon pump efficiencies, filter conditions, etc. They are generally not responsive to feedback sensors but are established during setup by the laboratory technician.

These control functions may be easily handled by software from the mainframe or a personal computer. However in many cases they are entered into a simple microprocessor by the technician. The mainframe or personal computer may then simply provide a signal to start this sequence. Automated dissolution systems intended to interface with HPLC should offer both options, e.g., microprocessor control of operating parameters to be started on signal from computer software, or alternatively input bus to enable such commands to come from the computer. When sampling or other operating function involve robotics [10], more operating functions are generally reserved for the computer than with modular systems.

Software that interprets HPLC data is available, and it is more comprehensive than that for other analytical methods that may be used in dissolution testing. There is no question that software for a personal computer or a mainframe should handle the data reduction and report results of dissolution testing. The only benchmark should be the availability of raw data irrespective of its manipulation.

Data reduction programs are commercially available. One problem, however, is that most of these are proprietary softwares. The source codes may or may not be available. It has been the author's experience that every customer has minor software variation requirements to fit his particular problem. Software should thus always include accessibility to the source code for in-house modifications as necessary. The author is aware that this poses severe problems for software producers. The commercial answers are not yet available at this writing.

REFERENCES

1. B. J. Compton and O. N. Hinsvark, A review of automated dissolution testing, *Pharm. Tech., 10*: 28 (1986).
2. H. M. Abdou, T.M. Ast, and F. J. Cioffi, Semi-automated system for high pressure liquid chromatographic determination of dissolution rate of fludrocortisone tablets, *J. Pharm. Sci., 67*: 1397–1399.
3. T. M. Ast and H. M. Abdou, Analysis of Fludrocortisone acetate and its solid dosage forms by high performance liquid chromatography, *J. Pharm. Sci., 68*: 421–423 (1979).
4. A. J. Owen, The diode-array advantage in UV/visible spectroscopy, Hewlett-Packard Publ. #12-5954-8912, Palo Alto, Calif.

Application of HPLC to Dissolution Testing

5. Transdermal Delivery Systems—General Drug Release Standards, *Supplement 8*, pp. 3080-3082, *USP XXI/NF XVI*, U.S. Pharmacopoeial Convention, 12601 Twinbrook Parkway, Rockville, Md. (1988).
6. K. A. Walters, Percutaneous absorption and transdermal therapy, *Pharm. Tech., 10*: 30-34 (1986).
7. T. J. Franz, The finite dose technique as a valid in vitro model for the study of percutaneous absorption in man, *Curr. Probl. Dermatol., 7*: 58-68 (1978).
8. D. C. Cox, C. C. Doluglas, W. B. Furman, R. D. Kirchhoefer, J. W. Myrick, and C. E. Wells, Guidelines for dissolution testing, *Pharm. Tech. 2*: 40-53 (1978).
9. T. S. Savage and C. E. Wells, Automated sampling of in vitro dissolution medium: effect of sampling probes on dissolution rate of prednisone tablets, *J. Pharm. Sci., 71*: 670-672 (1982).
10. Fully Automated Dissolution Testing, Product Information, Zymark Corporation, Hopkinton, Mass.
11. Technical Bulletin—Sampling Systems, Hanson Research Corporation, Chatsworth, Calif.
12. Observations; Pan American Health Organization Seminar on Dissolution, University of Panama, Panama City, Panama (1985).
13. D. C. Cox and C. Wells, In vitro dissolution of digoxin tablets, internal document supplied on request, National Center for Drug Analysis, FDA, St. Louis, Mo.
14. F. Langenbucher, Use of flow-through apparatus for dissolution testing, *5me Journée d'Actualités Biopharmaceutique de Clermont-Ferrand*, France (1980).
15. Sample changing chemical analysis method and apparatus, U.S. Patent 4, 108, 602, Hanson Research Corp. Chatsworth Calif. (1978).
16. Technical notes #7, Rheodyne Corporation, Cotati, Calif. (1986).
17. Wisp Autoinjector, Waters Associates, Milford, Mass. (1983).
18. Wisp 72D Dissolution System, Waters Associates, Milford, Mass. (1988).

9
Robotic Automation of HPLC Laboratories

Robin A. Felder

University of Virginia, Health Sciences Center, Charlottesville, Virginia

INTRODUCTION

High performance liquid chromatography (HPLC) is a technique commonly employed in a variety of laboratories. Because of its inherent flexibility, HPLC is suited to a wider range of analytical separation problems than any other single analytical method. HPLC has been a major analytical technique in the pharmaceutical industry, in clinical laboratories, and in commercial environmental laboratories [1]. Through the use of interchangeable hardware and an endless variety of separation media, almost any type of chromatographic separation is possible. Although HPLC is capable of rapidly separating many different substances with high resolution, it suffers from being a labor intensive technique.

The desire to incorporate robotics into the HPLC laboratory has been due in part to the tedious nature of the preparative workup. There are many successful examples of automating the autoinjection and data reduction steps in HPLC analysis. Sample preparation, however has lagged behind in becoming fully automated. The use of programmable automation, or robots, promises to finally provide the HPLC laboratory with the hardware flexibility necessary to eliminate many of the tedious steps necessary to prepare samples for chromatography. The inherent programmability of robots and the relative ease of changing the hardware configuration allow them to adapt to the constant improvements in the chemistry of sample preparation.

Robotic arms are precision mechanical devices that enable microcomputers to perform physical labor, which can be of great benefit to almost all laboratories. Generally, robots cannot perform tasks any faster than the laboratory staff, but results are more accurate, consistent, and reliable than those produced manually. Although robots require a large initial investment, not only financially, but in employee time, they quickly become economically feasible because they do not take breaks, can work unattended, and can perform tasks late into the evening without showing signs of fatigue. Unlike personal computers, which can quickly become outmoded because of insufficient memory or speed, robots have a long life expectancy. Mechanical devices are not as frequently outmoded and are less prone to technological obsolescence.

Robots are already in widespread use in industrial HPLC laboratories. It has been estimated that between 1000 and 1500 robots are in use in the approximately 75,000 industrial analytical laboratories in the USA [2]. Relatively recently, several turnkey robotic systems equipped with computers and analytical hardware and software, as well as robotic arms, have been introduced specifically for the automation of HPLC analysis. In this chapter we will be examining the state of the art in robotic hardware and its use in HPLC. Through examples, it will be shown that robotics is within reach of any laboratory both technologically and economically and is no longer the technology of the future but one that may be exploited today.

DEFINITION OF A ROBOT

Laboratory robots can take many forms. By definition, a robot is a machine that can be programmed to perform a task with humanlike skill [3]. Practically, the term robotics refers to programmable manipulators that perform a variety of skilled actions using a combination of mechanical and electronic components. All other devices designed for only one repetitive task or with limited mechanical capabilities and restricted programmability are referred to as "hard automation." For example, HPLC autosamplers are considered hard automation even though they are often called robotic autosamplers. The recent introduction of programmable pipetting devices with a high degree of programmability and movement within a Cartesian plane have come close to being robots except that they are not capable of mechanical manipulation and so by definition are not robots. They have been used for a variety of tasks such as solid phase extraction and will be discussed later.

Modern laboratory robots evolved from their larger ancestors, the industrial robot. The first robots were designed to handle superhuman tasks such as lifting heavy objects or rapid, highly accurate welding. It became apparent to robot manufacturers that there were many tasks in the laboratory that could be automated by a lighter duty robot. Although modern laboratory robots are simple to program and operate, there is considerable complexity behind the user friendly facade and diminutive size. Robots consist of three components: manipulators, control electronics, and software. All three are necessary for a minimally reprogrammable device, but many additional features may be added to enhance the basic robot such as sensors and high level programming languages.

The working robot is a complex interplay between the various motors controlling joints and the computer that drives them. Establishing robot motion control that mimics the smooth movement of the human arm and has a high degree of repositional precision is a difficult problem that has been solved using the science of kinematics [4]. Kinematics is applied to the robot in three levels of complexity. First, trajectory planning determines position, velocity, and acceleration for each movement made by the robotic manipulators. Secondly, inverse kinematics is applied to translate between movement required in the coordinate system and the particular geometry of the robot being developed. Finally, inverse dynamic equations are applied to establish how the robot moves under a variety of applied torques and forces. Each movement of the robot is represented, therefore, by a set of remarkably complex equations, the implementation of which has fortunately been simplified through the use of high level computer languages.

The tasks performed by a robot are usually defined by their complexity in terms of the degrees of freedom necessary to achieve certain orientations by the robot manipulators. Six coordinates are needed to quantitate the orientation of an object in space: x, y, and z coordinates locate the object in space, while three angles describe its orientation around its center of gravity. Robots can be categorized by the number of degrees of freedom and can move in straight lines (prismatic joints) in three dimensions defined by mutually orthogonal x, y, and z axes (Figure 1).

Cylindrical robots combine one rotational movement with two prismatic joints. Jointed or spherical robots usually have two rotational joints (waist and shoulder) and one prismatic joint. An elbow robot has three rotational joints (waist, shoulder, and elbow). Finally, a fully articulating robot has four rotational joints (waist, shoulder, elbow, and wrist) and is the most versatile robot manufactured commercially. SCERA robots have a unique double cylindrical movement that makes them unusually rapid in their execution of horizontal movement. Usually, additional degrees of freedom are obtained by the addition of a servo gripper, or hand. Even with three joints, a fully articulating robot does not come close to the 22 degrees of freedom found in the human arm.

Robot purchasers should match the complexity of the task with the levels of movement attainable by the robot. Robot flexibility or degrees of freedom can often determine whether or not a robot will be able to access an ergonometrically designed instrument. Most robot purchasers buy robots that are overachievers, and therefore they rarely use the versatility that can be attained by their instrument. Although there are many robots available suitable for the laboratory environment, several manufacturers have achieved market dominance through their aggressive marketing. Purchasing a robot can be difficult because there is no standard for quantitating robot performance [5]. Factors to consider when purchasing a robot for HPLC applications are speed, repeatability, accuracy, software, interfaceability, and available subsystems. Robotic performance and available peripherals have been compared by Beugelsdijk and Keddy [6].

AVAILABILITY

Laboratory robotic arms are available directly from the manufacturer. Many hours of simple instruction are required to educate the robot for even the simplest task.

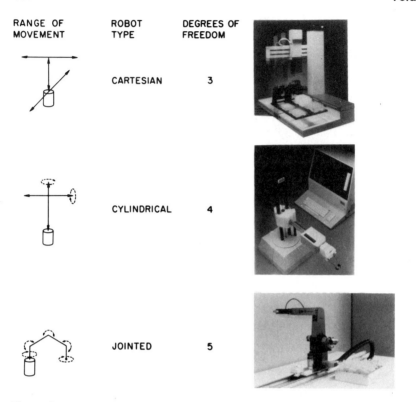

Figure 1 Three basic configurations of robots are used in research laboratories. A cartesian robot (top) has been used as the basis for many laboratory pipetting devices. Cylindrical robots (middle) move in a cylindrical space. Jointed robots (bottom) are used when the greatest degree of versatility is required.

Alternatively, a robot can be obtained as a system with a high level robot language as an added feature. Robotic systems are more user friendly since they can be programmed with simple English language instructions. Complete robot systems may also be purchased that are combined with many other useful automated devices to perform a specific task. Turnkey robotic systems are also available that are easy to operate and require no programming, but the initial capital outlay is substantial (often in excess of $60,000 for simple systems). Choices among laboratory robots become ever more complex as additional vendors offer robotic devices each year. Because of the volatility of the fledgling robot market, many vendors are also leaving the market place, so the stability of the robot vendor should be ascertained for continued service and support. Many of the robot vendors are based in foreign countries, making the availability of parts and service subject to the intricacies of the foreign markets and import restrictions.

APPLYING ROBOTICS TO HPLC

Many types of HPLC analysis have been automated successfully with robotics analysis for pharmaceuticals [7], drug testing in biological fluids [8], pesticide analysis [9], preparation and analysis of hazardous substances [10], and food component analysis [11, 12], to name a only a few. The impetus for considering robotics usually stems from large workloads, hazardous substances, or a need for increased precision. Each step of an HPLC analysis is an automation problem that can be solved either by the robotic arm itself or by the use of a dedicated automated peripheral device with which the robot may interact. Since robotic systems involve a complex interplay between a variety of peripheral devices, it is best to consider automating an HPLC procedure by first choosing the auxiliary equipment. The first step in choosing peripheral equipment is to break an HPLC procedure down into laboratory unit operations [13]. The manipulative steps of HPLC can be broken down into five basic laboratory unit operations (LUO) [13]: (a) obtaining the sample, (b) extraction or processing (solid or liquid phase), (c) storage, (d) injection via autosampler or direct injection, and (e) data reduction.

The many LUOs of HPLC require custom-made devices that can perform specific tasks. Zymark Corporation has pioneered the use of dedicated hands and devices for performing HPLC analysis. In fact, a patent has been issued to Zymark for their robot that can be programmed to change hands to accomplish different tasks. An HPLC robotic system may initiate an analysis with a hand designed to grasp test tubes for liquid extraction and finish the analysis with a hand specifically designed to inject specimens into an HPLC injection port. Peripheral devices are also available from Perkin Elmer Corporation, which was formerly in the robotics business.

The design of peripheral devices usually must allow for the limitations of movement and feedback control afforded by laboratory robots. Often there are design limitations in existing laboratory hardware that prevent its use by robots that do not have the dexterity of a human hand. A task such as opening a centrifuge or holding a tube on a vortex mixer is usually best accomplished with the use of a peripheral device designed with a robot in mind. Alternate task performance [14] is the subject of the following descriptions of robotics peripherals.

Autoinjectors

Automated devices have been developed that are designed to perform HPLC injections, relieving the technologist of the tedium of attending to the chromatography system. The laboratory robot is an alternative to the expensive autosampler, since it will inject each sample as it is extracted and processed. Although the laboratory robot may seem like an expensive autosampler (about twice the cost), when one realized that the robotic arm can process samples as well as inject them the costs can quickly become justified.

Three approaches to autoinjection have been used. The robot may be equipped with a dedicated injection hand designed to aspirate the sample followed by injection through the septum of the HPLC (Figure 2a). A dedicated HPLC injector is sold by Zymark that can process one sample at a time but has no sample storage capabil-

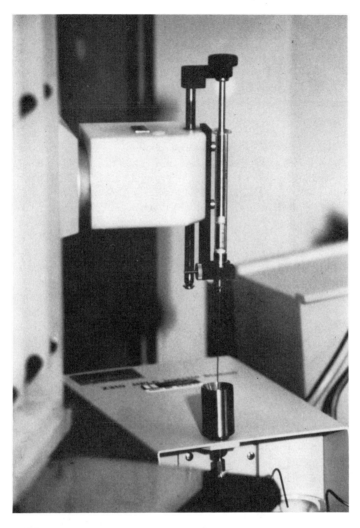

(a)

Figure 2 Three types of robotic autoinjectors have been configured for robotic operation. A Zymark robot can retrieve an injector hand that injects directly into the HPLC sample (a) Dedicated injection devices allow the HPLC equipment to reside outside the work area. (b) Samples may be aspirated directly from the sample tube in a sipping injector (c).

Robotic Automation of HPLC Laboratories

(b)

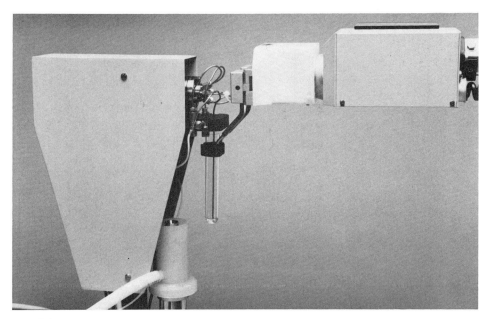

(c)

ity (Figure 2b). The dedicated injection device is equipped with a single port that may be accessed by a robot hand equipped with a syringe-holding and injecting device. The injector is also equipped with a 20 uL sample loop and tubing to connect it to most commercial HPLC pumps and chromatography columns. Zymark also manufactures a sipping injector for procedures that do not lend themselves to injection (Figure 2c). Autoinjectors allow the sampling end of an HPLC to exist within the performance envelope (the area that may be accessed) of the robot, while the HPLC itself may reside further away.

An alternate approach to sample handling is to use a robot to load an autosampler. In many models of autosamplers an articulating robot is necessary to place samples in the rather inaccessible sample tray. The advantage of an autosampler equipped with a sample tray is the ability to queue samples, freeing the robot to perform other tasks. The use of an autosampler is preferred when there is a large difference between chromatographic analysis time and sample preparation. Short analytical run samples may be prepared in advance of injection and be ready when the previous run is complete. Alternately, several robots may be used to keep pace with the chromatography. For long chromatography runs, the robot may be allowed to complete other productive tasks unrelated to the analysis. The greatest impediment to using a robot to load an autosampler is the interfacing necessary for robot computer management of the autosampler. Whittaker [15] demonstrated the use of a microcomputer to control an autosampler, for example rotating the sample tray to present an empty position to the robot. The autosampler and computer also work in conjunction with a cylindrical robot to form a complete integrated HPLC injector with a high degree of flexibility.

Extraction

HPLC requires a sample relatively free of background impurities that can interfere with the identification of chromatographic signals. Both liquid/liquid and liquid/solid phase extractions are used to partition the analyte of interest away from impurities while a washing step removes the interfering substances. The analyte of interest is then eluted either nonspecifically or selectively into a solvent that can be injected on the chromatograph. Sample extraction constitutes one of the most labor intensive steps in HPLC analysis. Furthermore, many of the solvents used in liquid/liquid extraction are noxious and require the use of a fume hood to protect the technologists from their harmful effects.

Robotics hardware has been recently developed that can perform almost any configuration of liquid/liquid phase or solid/liquid extraction. Many choices of solid phase separation material exist for HPLC including cationic, anionic, chelating, reverse-phase, and affinity matrices. The optimal separation material is often found after an initial educated guess followed by empirical trial and error. Johnson et al. [16] demonstrated that a Zymark robot and several custom-made devices could be used to alter the choice of solid phase packing material in separation columns until optimum conditions were met or it was determined that the extraction could not be optimized using the available reagents. Zymark offers a device that will dispense up

Robotic Automation of HPLC Laboratories 193

to 105 solid phase extraction columns through five separate dispensers (Figure 3). Operator definable variables include column condition solvent and time, extraction volume, extraction solvents, and elution time. Conceivably this would allow five extraction procedures to be run on a continual basis until columns were exhausted. Positive pressure is used to force the solvents through the column matrix after a grommet-equipped pneumatic cap covers the column.

One disadvantage of solid phase extraction by robot is that the extractions are performed serially. For chromatographic separations that are relatively slow, the robot can extract samples in time for the next injection. In the case of rapid chromatography, sample preparation becomes rate limiting.

Liquid extraction procedures involve more mechanical devices and more robotic steps than solid phase extraction. Zymark also offers liquid/liquid extraction peripherals, beginning with test tube dispensers that can accommodate up to 350 test tubes

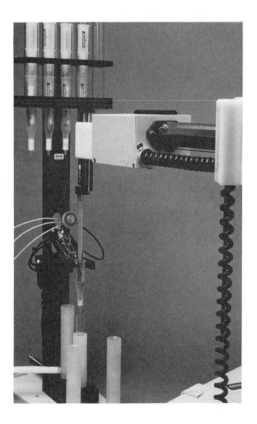

Figure 3 Solid phase extraction columns are dispensed in Dixie Cup style from a storage rack. Five types of extraction columns are available to the robot.

throughout the remainder of the extraction process. Either the dilute and dissolve station (Figure 4a) or the pipetting station (Figure 4b) may be used to dispense extraction solvent. Extraction solvent may be added in a fixed volume with the pipetting hand, or dispensed in a variable amount based on the volume of liquid preexisting in the extraction tube. This latter amount may be determined by preweighing the test tube (Figure 4b). After the extraction solvents are added, the extraction may proceed by vortexing in either a robot-compatible vortex (pictured in Figure 4a) or a tumble mixer (Figure 5). Phase separation may first require centrifugation in a robot-compatible centrifuge equipped with a pneumatic door and automatic indexing (Figure 6). Automatic indexing allows the first or index sample to be moved back to the same position each time in order to be located by the robot. Removal of either the

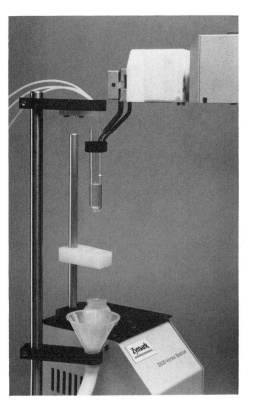
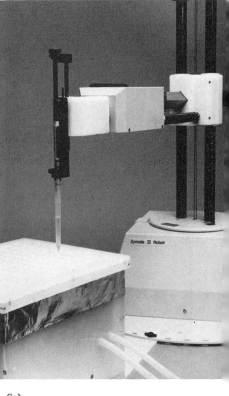

(a)   (b)

Figure 4 There are two basic approaches to the addition of organic solvents to allow liquid extraction by the robot. The sample may be brought to a specialized "dilute and dissolve" station that is under computer control (a). Alternately, the robot may add solvent at a pipetting station using a hand equipped with a pipette. (b).

Robotic Automation of HPLC Laboratories 195

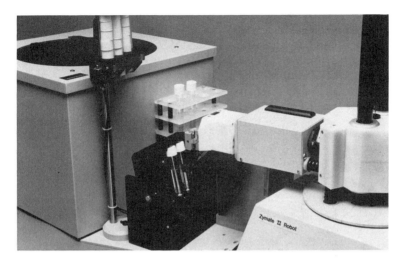

Figure 5 Sample extraction or mixing may be accomplished in a tumble mixer which can be loaded and unloaded by a robot.

Figure 6 A centrifuge engineered to allow the index position to return to the same location after each spin cycle prevents confusion over sample identity. Centrifuges may also be equipped with pneumatic lids to prevent aerosols from escaping into the laboratory.

upper or the lower layer may be accomplished with a liquid extractor station. This device allows one to choose the layer to be aspirated, the volume of aspirate, and the wash volume between extractions. Evaporation of unwanted solvent may be performed using either a heating mantle equipped with gas cannulae or a new Zymark evaporator that monitors solvent volume so that evaporation may be stopped at any time (Figure 7). The new Zymark evaporator is not currently robot compatible but should be in the near future.

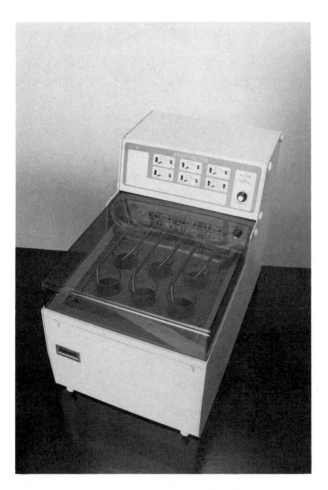

Figure 7 Sample evaporation can now be monitored and stopped at a predetermined level of dryness with a Zymark evaporator. Although not currently robot compatible, future models should allow robot access.

Robotic Automation of HPLC Laboratories 197

Additional Peripherals

Other necessary preparative steps prior to HPLC may be accommodated with additional peripheral equipment. A scale equipped with a pneumatic door may be interfaced to the robot controller so that fixed weights of substances may be added to test tubes (Figure 8). Particulate-containing sample may require filtration prior to analysis. A robot hand equipped with a pipette dispenses liquid into filtration unit that it had previously retrieved from a filtration unit dispenser. Positive air pressure is then used to displace the specimen through the filtration unit (Figure 9).

Many laboratories that venture into robotics projects fail to take into account either the cost of peripheral equipment or the time necessary to install, troubleshoot, and program these devices, although peripheral devices are necessary components to a successful HPLC robotics venture.

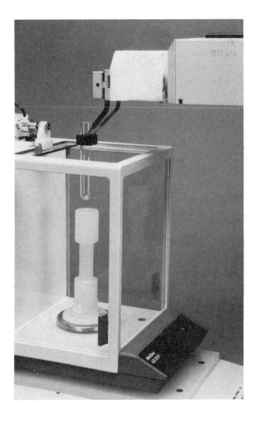

Figure 8 A laboratory balance equipped with a electronic interface allows precise solvent addition when combined with a pipetting hand. Monitoring the weight of solvent added with the host computer results in precise dilutions of sample.

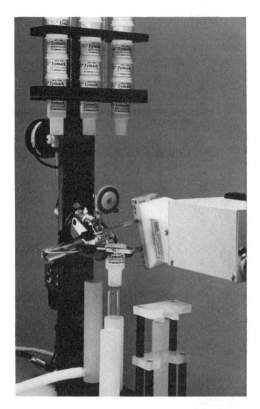

Figure 9 A filtration unit dispenser may either store many filtration units in any of three dispensers or allow a choice of one of three filter types. After the sample has been added to the filter unit, a hinged cap with a rubber sealing ring seals the top of the unit, followed by a positive pressure displacement of the liquid through the filter. The eluvate is either collected or aspirated to waste.

Throughput

A major concern of HPLC analyses is the number of samples that can be processed over a given period of time. The analysis of specimens by HPLC is usually accomplished serially and therefore is dependent on the separation time allowed by the chromatographic step. In many laboratories, the turnaround time is not critical except that it must be accomplished by the following morning. Robotics is often ideal for unattended overnight operation. Other investigators that have had requirements for rapid response time have used two HPLCs to double their output [17].

Alternately, the needs of a laboratory may include a variety of HPLC analyses over a short period of time. In this case the changing of columns and solvents requires constant attention by a trained technologist. In the robotics system designed by Antwerp and Venteicher [18], an automated switching device allowed the choice of any one of five HPLC columns. Furthermore, a solvent selector coupled with a programmable series of valves allowed selection of the appropriate solvent. The HPLC hardware was then followed by a programmable detector selector to allow automatic switching between UV or fluorescence. The reliability of this complex system was not discussed in terms of mean analytical time between failures.

ROBOTIC SYSTEMS
Cylindrical Robots

Many of the examples of HPLC systems coupled to robots employ Zymark robots. This is because Zymark is currently the only robot manufacturer that offers the robot, peripheral devices, customer support and training, and user friendly software. The cylindrical robot has been used in the clinical chemistry lab at the Cleveland Clinic Foundation in a high pressure liquid chromatography method that requires a complex series of steps involving sample extraction, separation of liquid phases, and injection [19] (Figure 10). Pippinger incorporated several Zymark robotic systems into a laboratory performing analysis of antidepressants. Medical technologists are necessary to prepare reagents and to place necessary supplies at designated

Figure 10 The chemistry laboratory at the Cleveland Clinic Foundation performs tricyclic antidepressant assays by robot. The robot laboratory is partially enclosed in a fume hood to protect laboratory workers against organic vapors.

locations within reach of the robot and to evaluate the quality of the final results. The robotic laboratory was placed in the opening of a fume hood to eliminate toxic fumes originating from extracted samples during the evaporation process.

The robot completes the drug extractions and sample injections using a specially designed injection hand into the chromatograph. For several years these robots have been performing their repetitive tasks with only minor malfunctions.

The cylindrical robot works in a cylindrical performance envelope. The four degrees of freedom exhibited by cylindrical robots (base rotation, elevation, movement in and out of a plane, and wrist roll) are usually sufficient for most HPLC laboratory operations. The major limitation of these robots is the lack of wrist pitch which can be useful for getting in and out of tight places. Additional flexibility in task performance is obtained by programming the robot to use a series of interchangeable hands.

A novel approach to engineering flexibility in the robotic laboratory, PyTechnology™, was invented by Zymark. PyTechnology, as the name implies, involves pie shaped table sections bolted to the circular table to which the robot is anchored (Figure 11).

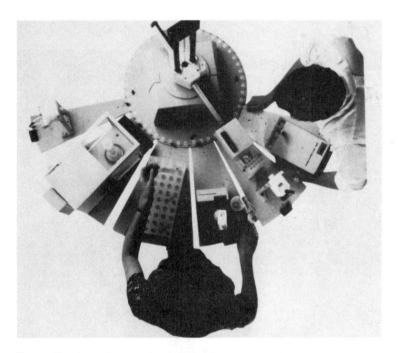

Figure 11 A novel approach to robotic laboratory organization is PyTechnology™. Each pie shaped table segment contains a peripheral device. Peripheral devices are quickly added in an orientation familiar to the robot, significantly reducing robot training time.

Each pie shaped wedge may contain a separate peripheral device, or several identical peripheral devices may be duplicated around one robot. The advantage of these wedges is that the location of the peripheral device on the pie is known by the robot, obviating the need for the robot trainer to teach the position of the device to the robot. What is needed is an identification of the pie device with which the robot must interact and the action desired, the rest is controlled by preprogrammed software. Like LUOS, the pie technology is a method of standardizing the hardware interface to the cylindrical robot. When procedures change, then new pie sections may be installed, and the action desired may be included in the user program. There are many recent examples of use of PyTechnology in HPLC procedures [17, 20- 23]. Most of these procedures involve the investigation of drugs in biological fluids because of the great utility in the pharmaceutical industry.

Articulating Robots

The most versatile robot available to the laboratory is the articulating robot. A recent example of a sophisticated articulating robot is the CRS robot marketed from Cyberfluor Inc. (Toronto, Canada) and the Master Lab robot formerly from Perkin Elmer Corp. (Norwalk, Conn.). Both robots have high degrees of flexibility with 5 degrees of freedom (Figure 1). The robot has shoulder, elbow and wrist joints rotating on a pivoting base. Furthermore, the robots have wrist pitch and roll maneuvers that allow access to areas often difficult to reach on analytical instruments. Positional accuracy of 0.5 mm or better is obtained by optically encoded discs, which must be set by nesting to a home or zero location each time the robot is turned on.

There are few examples of the application of articulating robots to HPLC. Schmidt and Dong [24] described a Perkin Elmer Master Lab Station robot for sample preparation and subsequent HPLC injection. Applications of this hardware configuration have been published by Dong [25] for the analysis of prednisone in tablets. The robotic procedure combined complete analysis and microcomputer data reduction that indicated whether the group of ten tablets passed or failed the predefined quality control tests. Precolumn derivitization of specimens was performed by an articulating robot in a procedure described by Dong [26], and Schmidt and Salit [27]. Precolumn derivitization is necessary for gas chromatography and involves the use of hazardous chemicals; therefore it is an ideal procedure for robotic automation.

Robotic Locomotion

For integrated HPLC laboratory design, the issue of sample delivery to the robot device must be considered. Several attempts at robot locomotion have been tried in the clinical setting where large numbers of samples arrive during a short period of time. Sasaki at Kochi Medical School has designed a laboratory in which patient specimens are transported to the robotics system via conveyor belt [28]. Samples are placed in numbered racks, which in turn are placed on conveyor belts at the front end of the laboratory, which transport them to the site of analysis. There are many belts that traverse the length of the laboratory, each delivering specimens to a specific analytical site; for example, all hematology tests are deposited on one belt. This method of locomotion allows the samples to be transported while the robots remain fixed in the center of a table to sample the specimens as they arrive into the robot area.

Another type of locomotion is shown by the computer-driven vehicles that move about hospital corridors picking up specimens and delivering them to the main laboratory [29]. Similarly, Sasaki [30] has designed robotic vehicles that move about the laboratory returning empty specimen racks to a central specimen receiving area.

Robots with Sensors

The mechanical performance of the robot can be enhanced by the addition of sensor technology on the hands or joints of the robot. A variety of mechanical and electronic sensor systems may be used. For example, computerized imaging systems might be used to check for sample integrity and container positioning for access by a robot. A simple vision system has been developed by Peck et al. [31] that employs a fiber optic device to sense the presence of a rubber septum in a specimen holding rack. The device allows precise placement of a cannula to puncture a septum based on the differential reflectance of the septum vs. the space between septa. Evasive action such as artery or system shutdown may be initiated by the misaligned robot, preventing hazardous solvent spills. Currently, video systems allow the greatest degree of intelligence to be imparted to a robot [32]. Several investigators are looking at the feasibility of tactile sensing in the fingertips of robot fingers. Tactile sensing approaching that of the human finger is in the foreseeable future [33].

The advantage of sensor technology is the ability of the robot to respond to changes in the analytical method. Analytical data can be checked by the robot host computer equipped with an expert system, and corrective measures such as sample reanalysis can be initiated if necessary. Many of these enhancements to increase the intelligence of the robotic system have not been examined in the HPLC laboratory setting. However, both Zymark and Cyberfluor robots have fingers that can sense the presence or absence of objects in their grasp. This feature is helpful when test tubes or syringes are dropped inadvertently during a procedure.

Power Requirements in the Robot Lab

One of the least discussed aspects of a robotic laboratory is the power supply, until power fluctuations result in a disaster. Power supplies to the robot should be filtered or under battery control if wide fluctuations are seen in the robot facility. Unattended robotics laboratories should be under battery backup power. If a robot handling toxic substances becomes disoriented as a result of a power failure then it could contaminate the laboratory by spilling the substance. Although the cost of clean power is often quite high, the costs of several power-related robot disasters often will cover the costs of an uninterruptable power supply.

CONTROL OF THE ROBOTS

Programming the Laboratory Robot

An important factor that has stimulated the introduction of robotics into the laboratory has been the development of high level robot programming languages coupled with interface technology. New robot languages offer simple robot programming through the use of English language commands. For example, the simple command GOTO MIXER initiates an intricate sequence of steps to drive the robotic arm to the mixing device. Several layers removed from the user's software, the computer generates electronic signals to the robot motion control mechanism to coordinate a smooth movement arc terminating at a precise location near the mixer. Complex algorithms employing robot kinematics translate computer machine code into signals that control acceleration upon commencing the movement and deceleration prior to stopping at the mixer. Furthermore, the robotic fingers are held parallel to the work surface throughout the complex series of movements to avoid spilling any liquid. Elaborate procedures can be developed by combining a series of simple commands that are programmed and tested individually. The robot can be instructed to pause in a procedure and examine the status of a sensor or instrument and then proceed through a choice of subsequent programs depending on the outcome of the test. Programmed intelligence of this sort can allow almost humanlike performance for many assays.

The integration of input and output ports with the robotic system is through interface technology. At the top of the control ladder is the robot-controlling software, consisting of a high level robot language and specific drivers for the low level language that drives the robot controller directly. An example of this hierarchical type of software system is shown in Figure 12. Each robot joint is controlled and coordinated through the robot controller language. Peripheral devices are regulated by the high level robot language through electronic interfaces. Zymark uses a dedicated device driver to interact with its peripherals, whereas most other robot manufacturers use circuit boards that plug into the standard IBM-PC bus.

Future robotics software development could be directed toward standardization and modularization of the basic operations performed in the laboratory [34]: sample manipulation, liquid handling, separation, conditioning, weighing, measuring, reporting, and storing. With a modular approach, high level robot control languages will reduce the time necessary for assay automation. Modular programming will allow rapid integration of several basic operation modules into a complete assay procedure with appropriate instrumental status checks. Standardization of peripheral hardware (i.e., centrifuge, mixer, and pipetter) interfaces will be essential for the rapid incorporation of various sample manipulations in the development of robotically controlled assays.

Artificial Intelligence

There have been numerous applications of artificial intelligence to the control of robotic HPLC applications. While many of the applications have not employed arti-

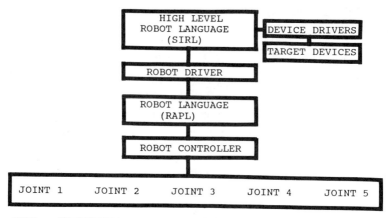

SIRL = STRUCTURED INTERACTIVE ROBOT LANGUAGE
RAPL = ROBOT AUTOMATION PROGRAMMING LANGUAGE

Figure 12 The software that controls a robot, peripheral devices, data management, and interfaces often exists in many levels of complexity and is written in a variety of computer languages. After the proper drivers are installed, most robot programming takes place in a high level robot language like SIRL (Structured Interactive Robot Language).

ficial intelligence as it is used by computer scientists (i.e., dedicated artificial intelligence languages such as Rulemaster), many investigators have used feedback control to determine whether an HPLC detector baseline was adequate for the analysis of pharmaceutical tablets. Only when the baseline was adequate did the analytic run commence. Furthermore, the system detected any malfunctions and shut the HPLC down in an orderly fashion. Li et al [36] have shown that a robot can be used to optimize solid phase extraction columns using simple if-then-else statements.

Expert systems have also been investigated in conjunction with robotics. Expert systems will ease the problem of maintaining and updating large knowledge bases as compared with conventional if-then-else programming. Lung and Lochmuller described the possibility of using expert systems in laboratory robotic systems. Lochmuller and Lung [38] went on to describe the use of feedback control of a robot to optimize colorimetric assays using a Simplex algorithm. A major chemical company is using a robot to optimize industrial chemical processes [39]. The robot will be used to examine one variable at a time until the optimum conditions for batch production of chemicals is found. According to Weglarz and Morabito [39], many industrial chemical processes may not currently be optimized because of the labor necessary to perform these tedious repetitive procedures. Artificial intelligence will someday be used to allow robotic systems to determine their own analytical parameters and to self-program robotic movement based on the results of previous "knowledge" acquired by the robot.

Prior to HPLC analysis, the optimization of sample extraction by solid phase cartridges was a tedious and time-consuming process. Many different cartridges with different solid phases must be tried under a variety of extraction conditions. Johnson et al. demonstrated that a Zymark robot and several custom-made devices could be used to alter the composition of the elution buffers and the type of packing in the columns until optimum conditions were met or until it was determined that the extraction could not be optimized using the available reagents. In a system designed by Ross and Vivilecchia [22], feedback control was used to monitor analytical results. The robot was then allowed to modify the system, such as sample preparation and injection volume, based on predetermined algorithms to optimize sample analysis.

User Interfaces

How friendly is robot software? The programming of most robotic software is beyond the scope of anybody without computer science training. There has been little attempt to make the training of robots or the design of robotic assays an easy task. Many robot software developers have assumed that scientists will have a certain level of prior experience when they place an order for a robot. Since the presumed higher level of programming sophistication was not present in most facilities, robots did not sell, and several manufacturers are no longer in the robot business. The term "user interface" implies a software design that makes many of the complex codes for robot motion control, and data input and output, transparent to the user. Simple English language commands should be able to train a robot to perform any task within its mechanical capabilities. Perkin Elmer and Zymark have developed simple robotic control languages that are accessible to most laboratory technologists. Unfortunately, no robot vendor has simplified all aspects of robot implementation; in particular, the use of interfaces remains problematic.

Robots are trained either through a teaching pendant (a group of switches on a wire, similar to a remote control) or directly through the robot keyboard. The robot is positioned by the trainer at a certain location, and then the coordinate is entered into the computer via a switch or the press of a key on the keyboard. A second coordinate may then be entered in a similar manner. Using simple commands from the keyboard, the coordinates are then replayed, and the robot will move as instructed. Robots, being both blind and without tactile senses, will collide with any obstacles in the path between the two points. Trainers must include a third point in the robot program that will allow a collision free trajectory. A recent innovation in robotic training is the "limp mode" employed by the CRS robot marketed by Cyberfluor (Toronto, Canada). In this mode, a robot trainer can simply grasp the robot arm and move it to a location. A press of a button automatically enters the position into the robot software, and the motion will be repeated once the software routine is started.

At the University of Virginia we are investigating the programming of robots with the use of touch screens. Digitized images (for example, a picture of the robot and HPLC on the computer screen) should allow the trainer to point to destinations in the picture to which the robot will then physically move [40]. Some future prospects for robot training may include the coupling of touch screen technology with digitized images of the work surface. The monitor would display a picture of the robotics laboratory from a choice of perspectives (top view or side view). A trainer could then trace the path the robot would take during the execution of a procedure on the computer monitor. Imaginative methods to train robots will simplify and accelerate the programming of a new procedures. Intellibotics (Oxnard, Calif.) has pioneered the commercialization of the user friendly graphics interface for programming the laboratory robot. Through the use of screen images, robotic procedures may be designed on the computer screen and tested before spending time running the robot.

INTEGRATED APPROACHES TO ROBOTICS LABORATORY DESIGN
Networking of Multiple Robotics Laboratories

Networking of multiple robotics laboratories is being investigated at the University of Virginia [41]. Through a local area network design (LAN) data from the robot computer, providing status of robotic operation as well as analytical results, are gathered continually on a file server computer. A continually updated database of all specimens arriving into the laboratory is provided to each robotic area to alert it of samples to expect. Total integration of robots instruments and user interaction stations (entry port for samples into the system) will allow many robotics laboratories to be monitored by a single individual.

Many commercially available networks are available that include software and hardware devices. One computer is usually dedicated as a file server or traffic director. Data can be transferred into the network through many communications protocols, but two popular and time-tested systems are Ethernet and Token Ring. Either protocol will perform well in robotics laboratories, but both must be evaluated by taking into account the need for speed and the capacity to handle many computers.

Networked multiple robotics laboratories will be the most labor efficient method of automating multiple analytical laboratories in large industrial settings.

Multiple Robotics Systems

For high throughput operations, a combination of programmable mechanical devices may be used. Two robots working in tandem can pass work completed in one area to a second step in another area. At the University of Virginia we are investigating the use of a programmable pipetting station in combination with a robotic arm to increase laboratory efficiency [42]. The analysis of glycated hemoglobin is an example of solid phase extraction not unlike that used for preparation of samples for HPLC. Extraction was accomplished by automatic transfer of samples to extraction columns, washing, and then elution into spectrophotometer tubes using a custom-designed trolley system. After the eluates were collected, a robotic arm placed them into a spectrophotometer equipped with a sample changer and an automatic data reduction system. Results were comparable in manual and robotic procedures. While one analytical method is being processed by the pipetting station, the robot will be preparing reagents for the next analytical run. Total front to back automation is possible with the advantage of serialized batch throughput.

A commercial multiple automation system is available that combines a programmable pipetting station, a microplate reader, a laboratory balance, an incubator, and a robot into an integrated device (Synchron Instruments, Etten-luer, the Netherlands). A cylindrical robot on a linear track allows five degrees of freedom. Electronic communication is provided through multiplexed RS-232 interface boards residing in two microcomputers. This is the first commercialization of an integrated approach to robotics.

IMPACT OF ROBOTICS ON TECHNOLOGISTS

The challenges facing laboratory technologists today are not unlike those that occurred after the introduction of the laboratory computer. Today's laboratory technologists are more familiar with electronics and computers than those ten years ago, but the specific knowledge required for robotics implementation is not readily found. However, technologists in the HPLC laboratory have been well equipped to adapt to this emerging technology because of the technical knowledge required to assemble and operate HPLC hardware.

There is a need for training programs that will produce technologists at the M.S. or Ph.D level who can design and implement robotics systems for clinical laboratories. Introduction of laboratory robotics will further redefine the role of the technologist, emphasizing method development and trouble shooting and deemphasizing processing and delivering samples to analyzers. Many tasks that require the skills of technologists can be automated by robot, but the potential for undetected sample errors increases with the degree of automation. The creation of hardware capable of determining problems with the sample (i.e., nonhomogeneity, incorrect labelling, improper containers) will be one of the major challenges for designers of laboratory robotics. Technologists of the future must receive at least modicum of training in this emerging field to be competitive for existing jobs.

LABOR

One to two personnel full time equivalents (FTE) are required for each robotic method for the duration of the development and installation procedure. Several months are often required for trained personnel to complete each project. Management supervision is required to maintain the project's continuity and its perspective in relation to the general configuration of the laboratory. Once the method is producing data, personnel knowledgeable in the system design must be in place to repair or to modify procedures as needs arise. Personnel who were formerly involved in method development can then be used to develop additional methods. At some point, as the automated methods proliferate around the laboratory, there will be decreased labor needs. The time frame for this eventuality is usually sufficiently lengthy that attrition will reduce the labor force without the need for forced labor reductions.

CONCLUSION

The robotics revolution has clearly begun in the analytical laboratory. Fully automated HPLC laboratories are no longer curiosities; they exist in many pharmaceutical laboratories. The payback for most simple systems is less than a year after several months are dedicated to installation. Future developments in sensor technology and the use of expert systems software will result in systems that will begin their payback after only a minimum of programming. Networking of robotic HPLC devices to data analysis and billing computers will result in a fully automated laboratory with minimal personnel requirements. While many mundane chores in the laboratory will be eliminated, there will be increased need for analytical chemists to design new problems for robots to solve.

ACKNOWLEDGMENTS

We wish to thank Vicki Hodges for her expert assistance in printing the manuscript, the Perkin Elmer Corporation for a generous grant, and Brian Lightbody at the Zymark Corporation for permission to use several of their figures.

REFERENCES

1. K. Conroe and B. Bidlingmeyer, The growth of HPLC applications, *Amer. Lab.*: 82-87, Oct. (1988).
2. B. Fowler, Integrated laboratory automation, a workshell from LIMS to robotics, *Amer. Lab.*: 62, Sept., (1988).
3. Edward L. Safford, *Handbook of Advanced Robotics*, Tab Books, Blue Ridge Summit, Penn. (1982).
4. Y. Wang and S. Butner, Robot motion control, *AI Expert*: 26- 32, Dec. (1987).
5. W. A. Schmidt, J. J. Rollheiser, and K. M. Stelting, The laboratory robotics evolution, a survey of today's robot options, in *Advances in Laboratory Automation Robotics*, Vol. 6 (J. R. Strimaitis and G. L Hawk, eds.), Zymark Corp., Hopkinton, Mass. (1989).
6. T. J. Beugelsdijk and C. P. Keddy, A critical comparison of the Zymark and Perkin-Elmer laboratory robotic systems, *Advances in Laboratory Automation Robotics*, Vol 3 (J. R. Strimaitis and G. L. Hawk, eds.),Zymark Corp., Hopkinton, Mass., pp. 503-532 (1986).
7. W. A. Davidson and K. Lam, Robotic HPLC analysis of solid dosage forms, *Advances in Laboratory Automation Robotics*, Vol. 4 (J. R. Strimaitis and G. L. Hawk, eds.), Zymark Corp., Hopkinton, Mass., pp. 71-94 (1987).
8. G. F. Plummer, The automation of analytical methods for the determination of drugs in biological fluids, *Advances in Laboratory Automation Robotics*, Vol. 3 (J. R. Strimaitis and G. L. Hawk, eds.), Zymark Corp., Hopkinton, Mass., pp. 47-26 (1986).
9. I. Laws and R. N. Jones, Generic sample preparation system for automation of pesticide residue analysis, *Advances in Laboratory Automation Robotics*, Vol. 4 (J. R. Strimaitis and G. L. Hawk, eds.), Zymark Corp., Hopkinton, Mass., pp. 15-26 (1987).
10. J. W. Brodack M. R. Kilbourn, and M. J. Welch, Automated production of several positron-emitting radiopharmaceuticals using a single laboratory robot *Appl. Radiat. Isot. 39*: 689-698 (1988).

11. D. J. Higgs and J. T. Vanderslice, Application and flexibility of robotics in automating extraction methods for food samples, *J. Chromatogr. Sci.*, 25: 187-191 (1987).
12. M. Dulitzky, A robotic method of caffeine analysis, *Amer. Lab.*: 104-109, June (1986).
13. W. J. Hurst and J. W. Mortimer, Laboratory robotics: A guide to planning, programming, and applications, VCH, New York (1987).
14. B. E. Kropscott, L. B. Cayne, R. R. Dunlap, and P. W. Langvardt, Alternative task performance in robotics, *Am.Lab.*: 70- 75, June (1987).
15. J. W. Whittaker, Use of an IBM system 9000 computer and LC/9505 liquid chromatographic automatic sample handler with the zymate laboratory automation system, *Advances in Laboratory Automation Robotics*, Vol. 1 (J. R. Strimaitis and G. L. Hawk, eds.), Zymark Corp., Hopkinton, Mass., pp. 331-341 (1984).
16. E. L. Johnson, K. L. Hoffman, and L. A. Pachia, Automated sample analysis from method development to sample analysis, *Advances in Laboratory Automation Robotics*, Vol. 5 (J. R. Strimaitis and G. L. Hawk, eds.), Zymark Corp, Hopkinton, Mass., pp. 111-117 (1988).
17. J. R. Curran, Automation of the invermectin HPLC content uniformity assay for Heartgard 30 tablets using PyTechnology robotics. *Advances in Laboratory Automation Robotics*, Vol. 6 (J. R. Strimaitis and G. L. Hawk eds.), Zymark Corp., Hopkinton, Mass (1989).
18. J. V. Antwerp and R. F. Venteicher, Improving the flexibility of an analytical robotic system, *LC:GC*, 4(5): 458-460 (1985).
19. C. Pippinger, The robots are coming, *Med. Lab. Observer*, 30: 4-11 (1985).
20. D. O. Chryst, W. H. Hong, and R. E. Daly, Application of PyTechnology in pharmaceutical analysis, *Advances in Laboratory Automation Robotics*, Vol. 6 (J. R. Strimaitis and G. L. Hawk, eds.), Zymark Corp., Hopkinton, Mass. (1989).
21. S. Conder and J. Kirschbaum, Automated stability testing of tablet dosage forms development of a grinding pysecton, *Advances in Laboratory Automation Robotics*, Vol. 6 (J. R. Strimaitis and G. L. Hawk, eds.), Zymark Corp., Hopkinton, Mass. (1989).
22. B. Ross and R. V. Vivilecchia, Automated preformulation stability studies, *Advances in Laboratory Automation Robotics*, Vol. 6 (J. R. Strimaitis and G. L. Hawk, eds.), Zymark Corp., Hopkinton, Mass. (1989).
23. M. Chang, L. Kosobud, and G. Schoenhard, Development of a high throughput totally automated assay for SC-47111, new quiniline antibacterial agent, *Advances in Laboratory Automation Robotics*, Vol. 6 (J. R. Strimaitis and G. L. Hawk, eds.), Zymark Corp., Hopkinton, Mass. (1989).
24. G. J. Schmidt and M. W. Dong, Robots: Applications in automated sample analysis, *Am. Lab.*: 62-72, February (1987).
25. M. W. Dong, The use of the laboratory robotics for sample preparation, *J. Liq. Chromatogr.*, 9(14): 3063-3092 (1986).
26. M. W. Dong, Analytical derivatization via robotics: Amino acid analysis by precolumn Ophthaladehyde derivitization, *LC-GC*, 3: 255-260 (1987).
27. G. J. Schmidt and M. Salit, A general robotic procedure for performing precolumn derivitization and its application to on-line liquid chromatographic analysis, Perkin-Elmer Technical Bulletin 10, pp. 1-18 (1987).
28. M. Sasaki, The belt line system, completed automatic clinical laboratory by sample transportation system, *Jpn. J. Clin. Pathol.*, 32: 119-126 (1984).
29. D. Seligson, Robotics *Clin. Chem.*, 29: 1154 (1983).

30. M. Sasaki, An attempt for transporting of laboratory samples. The establishment and further development of the belt line system, *JJCLA, 10*: 82-90 (1985).
31. C. N. Peck, K. M. Lehr, and N. E. Lutenske, Automation of high performance liquid chromatography analyses using a Zymate laboratory robot, *Advances in Laboratory Automation Robotics*, Vol. 5 (J. R. Strimaitis and G. L. Hawk, eds.), Zymark Corp., Hopkinton, Mass., pp. 1-14 (1988).
32. W. M. Zuk, M. A. Perozzo, and K. B. Ward, Automated preparation of protein crystals: Integration of an automated visual inspection station, *Advances in Laboratory Automation Robotics*, Vol. 4 (J. R. Strimaitis and G. L. Hawk, eds.), Zymark Corp., Hopkinton, Mass. (1988).
33. Tactile sensing approaching that of the human finger is in the foreseeable future, *NASA Technical Briefs*, Vol. 11: 12-13 (1987).
34. H. M. Kingston and F. Ruegg, Computer-controlled robot system for column chromatography separations, *Advances in Laboratory Automation Robotics*, Vol. 6 (J. R. Strimaitis and G. L. Hawk, eds.), Zymark Corp., Hopkinton, Mass. (1989).
35. K. Halloran and H. Franze, Interaction between a robotic system and a liquid chromatograph: HPLC control, communication, and response, *LC-GC, 4*: 1020-1025 (1988).
36. C. Li, J. Potucek, and H. Edelstein, Automated method development of liquid solid extraction, *Advances in Laboratory Automation Robotics*, Vol 4 (J. R. Strimaitis and G. L. Hawk, eds.), Zymark Corp., Hopkinton, Mass., pp. 28-40 (1987).
37. K. R. Lung and C. H. Lochmuller, The anatomy and function of the laboratory robot, *J. Liq. Chromatogr. 9*: 2995-3031 (1986).
38. C. H. Lochmuller and K. R. Lung, Applications of laboratory robotics in spectrophotometric sample preparation and experimental optimization, *Anal. Chim. Acta, 183*: 257-262 (1986).
39. T. E. Weglarz and P. L. Morabito, The use of laboratory robotics for automated process optimization, *Advances in Laboratory Automation Robotics*, Vol. 6 (J. R. Strimaitis and G. L. Hawk, eds.), Zymark Corp., Hopkinton, Mass. (1989).
40. A. Martinez, D. P. Vaughn, K. S. Margrey, and R. A. Felder, Touchscreen/graphics interface technology for a remotely monitored, robotic clinical laboratory, *Advances in Laboratory Automation Robotics*, Vol. 6 (J. R. Strimaitis and G. L. Hawk, eds.), Zymark Corp., Hopkinton, Mass. (1989).
41. D. P. Vaughn, K. S. Margrey, A. Martinez, J. C. Boyd, J. Savory, and R. A. Felder, Software integration for a satellite robotic laboratory system, *Advances in Laboratory Automation Robotics*, Vol. 6 (J. R. Strimaitis and G. L. Hawk, eds.), Zymark Corp., Hopkinton, Mass. (1989).
42. R. A. Felder, W. D. Holman M. M. Wu, J. R. Spina, H. M. Barnwell, and A. martinez, Increasing robotic laboratory efficiency through the combination of pipetting station and robotic arm to perform two popular assays: Radioligand binding and solid phase extraction, *Advances in Laboratory Automation Robotics*, Vol. 6 (J. R. Strimaitis and G. L. Hawk, eds.), Zymark Corp., Hopkinton, Mass. (1989).

Part Four
HPLC of Peptides, Proteins, and Enantiomeric Drugs

10

Liquid Chromatographic Resolution of Enantiomers of Pharmaceutical Interest

Khanh H. Bui

ICI Pharmaceuticals Group, Wilmington, Delaware

INTRODUCTION AND BACKGROUND

Stereochemical analysis has become an increasingly important problem in the pharmaceutical field, since numerous pharmacologically active agents are chiral and their two enantiomeric forms usually exhibit different physiological properties. Indeed, numerous examples exist where the two enantiomeric forms manifest different pharmacological actions, potencies, biodistribution and disposition kinetics, and/or host toxicities. For instance, most of the β-blocking activity of β-blockers such as propranolol and metoprolol is attributed to the (S)-enantiomer [1-2]. Furthermore, stereoselective differences have also been observed in the metabolism and pharmacokinetics of the two enantiomers of the above mentioned compounds [2-5].

Traditional methods for determining the enantiomeric composition of chiral compounds, such as optical rotation measurement and/or fractional recrystallization of diastereomeric salts, are usually difficult, insensitive, inaccurate, and limited in applicability. Hence modern chromatography, because of its reproducibility, accuracy, selectivity, sensitivity, and speed, has become the technique of choice for the analysis of enantiomers. Since most of the recent developments and applications in enantiomeric resolution have been performed using liquid chromatography because of its wider applicability, particularly in the pharmaceutical field, this chapter will be restricted to liquid chromatography of enantiomers. Readers interested in gas chromatography of enantiomers are referred elsewhere [6,7].

Generally speaking, there are three different approaches under which chromatographic resolution of enantiomers can be achieved. The first approach involves the precolumn conversion of the enantiomers to diastereomers by reaction with a chiral derivatizing agent. The resulting diastereomers are then resolved on an achiral stationary phase using an achiral mobile phase. The second approach involves the direct resolution of enantiomers on an achiral stationary phase using a mobile phase that contains a chiral component. The third approach, which recently has received a great deal of research and development, involves the direct resolution of enantiomers using selective chiral stationary phases.

In this chapter, the chemistry, applicability, and limitations of the three different liquid chromatographic techniques for the resolution of enantiomers will be examined. Particular emphasis will be given to the chromatographic approach, in which chiral stationary phases are utilized, since this area has recently enjoyed numerous interesting developments.

INDIRECT LIQUID CHROMATOGRAPHIC RESOLUTION OF ENANTIOMERS VIA PRECOLUMN FORMATION OF DIASTEREOMERIC DERIVATIVES

Principle and Mechanism

Unlike enantiomers, whose physical properties are identical, diastereomers may have very different properties, such as solubility and polarity, that permit them to be easily resolved by ordinary chromatographic means. Therefore, by converting enantiomers to diastereomers via chemical reactions with chiral derivatizing reagents, their chromatographic separation can be easily achieved. Oftentimes, the most difficult task in this chromatographic approach involves the selection of a suitable chiral reagent rather than the development of a chromatographic procedure for separating the derivatized diastereomers. Indeed, there are several factors involved in selecting a chiral reagent that can affect the applicability, accuracy, and resolvability of the chromatographic procedure.

1. First, the chiral reagent should be stable and easy to prepare or commercially available in a state of high optical purity. This is important since the degree of optical purity of the chiral reagent has a direct influence on the accuracy and the maximum detectable optical purity of the enantiomeric solutes [5,8].
2. The reaction should be mild, so that no racemization of the chiral center of the solutes and reagent occurs.
3. The derivatization must be quantitative and complete. Otherwise, determination of the enantiomeric purity could be inaccurate, since the reaction of enantiomers with a chiral substrate can occur at different rates [9].
4. The chiral solute should contain only one functional group for derivatization.
5. To enhance the detectability of the derivatives, the chiral derivatizing reagent should contain a chromophore or fluorophore.

Table 1 A Summary of Chiral Derivatizing Reagents and Their Chromatographic Applications

Chiral reagents	Functional group reacted to	Separation mode	Applications	References
Isothiocyanates				
2,3,4,6-tetra-O-acetyl-β-D-glucopyranosyl	Amine	RP	β-Blockers, catecholamines, adrenergic agents, and other amino alcohols	12–15
R-α-methylbenzl		RP	Propranol, ephridrine	16
2,3,4-tri-O-acetyl-α-D-arabinopyranosl		RP	Catecholamines	13
Isocyanates	Amine			
S(−)-1-phenylethyl-			Propranolol	17
R(−)-1-(naphthyl)ethyl-			Oxprenolol	18, 19
Acyl chlorides	Amine			
N-trifluoroacetyl-S-prolyl-		RP	Propranolol	8
(phenyl sulfonyl) prolyl-			Amphetamine, ephedrine and pseudoephedrine	20
Anhydrides	Hydroxyl			
t-BOC-L-ala-		RP	Propranolol	21
t-BOC-L-leu-			Alprenolol	22
O,O derivates of (R,R) tartaric acid-			β-Blockers	5
Amines	Carboxylic acid			
D and L -1-(4-(dimethylamino-1-napthyl)-ethylamine		N	Ibuprofen, indoprofen, naproxen	21, 22
D and L-1-(1-anthryl)ethylamine		N	Naproxen	23
D and L-1-(2-anthryl)ethylamine			Naproxen	23
Acids	Amine			
(+) and (−)-α-methyl-1-napthaleneacetic acid		N	DOPA, amino acids, and ornithrine	24

RP – Reversed Phase; N – Normal Phase

6. Furthermore, since the chromatographic resolvability of the derivatized diastereomers is dependent on the degree of difference in various properties possessed by them, such as polarity, energies of adsorption, the accessibility of various polar substituents, and molecular structure [10,11], a derivatizing reagent that can maximize the above differences should be selected. Indeed, it has been demonstrated that the chromatographic resolution is greatly enhanced by the following factors: (a) the conformational ridigity of the groups attached to the chiral center of the derivatizing reagent, (b) the short distance between the two chiral centers of the derivatized diastereomers, and (c) the large size differential in the groups attached to the chiral reagent [5,11].

Applications

Recently reported chiral derivatizing reagents and their chromatographic applications for the analysis of enatiomeric drugs are listed in Table 1. Various chiral isothiocyanate, isocyanate, acyl chloride, and anhydride derivatives were used to derivatize β-adrenergic blocking drugs containing an aminol alcohol functional group, while several chiral amines were used to derivatize nonsterodial antiinflamatory drugs containing a 2-aryl propionic acid functional group. Several of the reagents listed exhibited remarkable generality in resolving enantiomers having similar functional groups. For example, the use of 2,3,4,6-tetra-O-acetyl-β-D-glucopyranosyl isothiocyanate and the 0,0 derivatives of (R, R) tartaric acid anhydrides brought about the enantiomeric resolution of a wide range of amino alcohols [5,12,15]. Nevertheless, chiral derivatizations of compounds of other classes still have to be investigated on a case by case basis. This and the unavailability of several chiral derivatizing reagents listed with high optical purity [8, 22-23] severely limit the use of this chromatographic approach for the analysis of enantiomers.

CHROMATOGRAPHIC RESOLUTION OF ENANTIOMERS USING CHIRAL MOBILE PHASES

This chromatographic approach, which requires no tedious precolumn derivatization of the sample and no specialized chiral stationary phase, involves the use of mobile phases containing novel chiral additives to resolve enantiomers. Based on the type of chiral additives and the mechanism under which the enantioselectivity is achieved, this approach can be classified into three classes: ligand exchange chromatography, chiral ion pair chromatography, and chromatography using mobile phases containing macromolecular aggregates.

Ligand Exchange Chromatography

The two general approaches in which ligand exchange chromatography is performed are illustrated in Figure 1. The first approach, pioneered by Davankov for the resolution of amino acid enantiomers, employed stationary phases consisting of

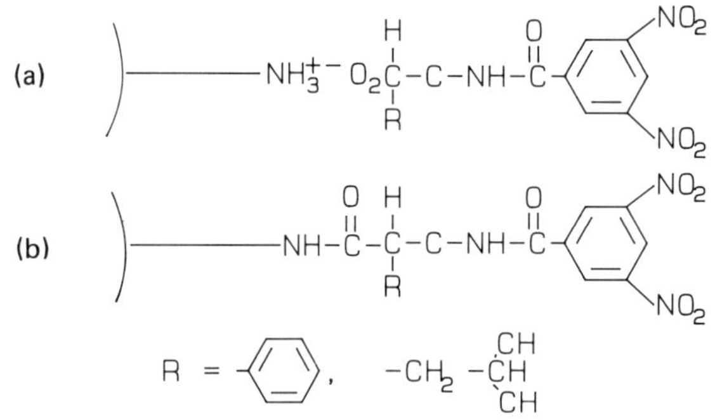

Liquid Chromatographic Resolution of Enantiomers

(a) }————NH₃⁺⁻ O₂C-C(H)(R)-C-NH-C(=O)-C₆H₃(NO₂)₂

(b) }————NH-C(=O)-C(H)(R)-C-NH-C(=O)-C₆H₃(NO₂)₂

R = –C₆H₅, –CH₂–, –CH(CH)(CH)

Figure 1 Schematic illustration of (a) bonded proline-copper amino acid complex and (b) dynamic proline-copper amino acid complex.

amino acid ligands bonded to polymeric resins on which Cu (II) metal ions were subsequently loaded [25–28]. These polymeric sorbents, although having exhibited remarkable enantioselectivity for amino acids, were not stable to pressure and hence not applicable to HPLC. Consequently, a series of chiral ligands was bonded to silica gels; under either normal or reversed phase mode, this resulted in very efficient enantiomeric separation of amino acids and amino acid derivatives [28–32]. To date, most of the applications using this approach have been centered around the enantiomeric analysis of amino acids, and no enantiomeric drug analysis has been reported [28,32].

The second approach, termed dynamic ligand exchange chromatography, was pioneered in 1979 by Karger and coworkers and by Hare and Gil-Av for the resolution of amino acid enantiomers [33–35]. They showed that by adding chiral metal complexing agents to the mobile phases, very efficient separation of amino acid enantiomers and their derivatives can be achieved using conventional achiral stationary phases such as reversed phase or ion exchange packings. The mechanism of the resolution is based on the difference in the stabilities and therefore the equilibrium concentrations as well as the difference in the partitioning and/or adsorption behaviors of the two diastereomeric ternary complexes formed by the solute enantiomers and the chiral metal ligand [27, 28].

Even though the bulk of the applications using the second approach also involved the resolution of amino acid enantiomers and their derivatives [28,32], its usefulness in resolving other compounds has been demonstrated. Various dynamic ligand exchange chromatographic resolutions of non-amino-acid compounds are illustrated in Table 2. It should be noted that in most cases, the metal complexing agent consists of a certain optically pure L-amino acid or derivative and Cu (II). Copper (II) was the most frequently used transition metal, probably because of its strong chelating ability [36].

Table 2 Enantioselective Dynamic Ligand Exchange Chromatography of Non-Amino-Acid Compounds

Chiral selector	Metal ion	Stationary phase	Solute enantiomers	References
L-proline	Cu^{2+}	Silica gel	Thyroid hormones α-Methyl DOPA	37
L-pro,L-val,L-ile,L-phe N-methyl-L-proline	Cu^{2+}	C_{18}	α-Hydroxy acids	39
L-proline	Cu^{+2}	C_{18}	Normetanephrine N-acetyl-D-L-penicillamine pipecolic acids	38
(R, R) - + -tartaric acid mono-n-octylamine	Cu^{+2}	C_{18}	Ethanolamines	36
L-asp-L-phe-o-CH_3	Cu^{+2}, Zn^{+2}	C_{18}	DOPA	40, 41

Chiral Ion Pair Chromatography

Ion pair chromatography is a technique in which the retention of ionized compounds is regulated by the addition of a counter ion into the mobile phase. If the counter ion is chiral, stereoselective separation can be achieved. The mechanism of the chiral separation is based on the difference in partitioning and/or adsorption behaviors of the diastereomeric ion pairs formed by the enantiomeric solutes and the chiral counter ion. Usually the binding forces consist of electrostatic and hydrogen bonding [41], and mobile phases of low polarity are normally used to allow a satisfactory degree of ion pair formation. The stereoselectivity of the separation is dependent on the structure of the counter ion. Usually the enantioselectivity of the separation is high when the counter ion contains binding groups having a rigid ring system with bulky groups in the vicinity of the chiral center [41]. Recent applications using this chromatographic technique are summarized in Table 3.

Chromatography Using Mobile Phases Containing Macromolecular Aggregates

Natural macromolecules such as proteins and cyclodextrins have been known to exhibit stereoselective binding with low molecular weight molecules [46-48]. Consequently, when macromolecules are used as mobile phase additives, stereoselective separations can sometimes be achieved. In general, the retention of the enantiomeric solutes as well as the selectivity of the separations are regulated by the concentration of the macromolecules. An increase in the concentration of the macromolecules usually decreases the retention of the solutes (by enhancing their partitioning to the mobile phase) and increases the selectivity of the separation [41,49-51]. The various stereoselective separations utilizing macromolecular additives reported in the literature are summarized in Table 4.

CHROMATOGRAPHIC RESOLUTION OF ENANTIOMERS USING CHIRAL STATIONARY PHASES

This chromatographic approach, which enjoys several advantages over the first two approaches, has been the major focus of interest in many laboratories over the past several years. Indeed, the use of chiral stationary phases not only permits the use of inexpensive ordinary mobile phases (which do not interfere with the detection of the enantiomers) but also obviates the need to derivatize the solutes with chiral reagents (which may be impure, may be unavailable, or may exhibit different reaction rates with the enantiomers).

The various chiral stationary phases reported can be classified as (a) Pirkle type chiral stationary phases, (b) cyclodextrin bonded stationary phases, (c) protein bonded stationary phases, (d) chiral polymer stationary phases, and (e) other chiral stationary phases. Several chiral stationary phases are now commercially available. Their trade names and manufacturers are listed in Table 5.

Table 3 Chiral Ion Pair Chromatographic Separations

Chiral counter ion	Enantiomer solutes	Mobile phase	Stationary phase	References
Quinine	Carboxylic acids (tropic, atropic naproxen, etc.)	CH_2Cl_2/Pentanol	Diol	42, 44
Camphorsulfonate	β-Blockers	CH_2Cl_2/Pentanol	Diol	43
Tartaric acid ester/ Hexafluorophosphate	Ephedrine and analogues	Phosphate buffer	Phenyl	45

Table 4 Enantiomeric Resolutions Using Mobile Phases Containing Macromolecular Aggregates

Mobile phase additives	Stationary phases	Solutes	References
Human serum albumin	C_{18}	Carboxylic acids	41
β-Cyclodextrin	C_{18}	Barbiturates, mandelic acids, mephenytoin	49–51
γ-Cyclodextrin	C_{18}, C_8, CN	Norgestrel	52
α-Cyclodextrin	C_{18}	Mandelic acids	51

Table 5 Commercially Available Chiral Stationary Phases

Types of chiral stationary phases	Trade name	Manufacturer
Pirkle type chiral stationary phases		
(R)-N-DNB-phenylglycine (ionic)*	Type I-A	Regis[a]
	DNBPG (ionic)	Baker[b]
	ChrilRsil-I	Alltech[c]
	Sumipax 0A2000	Sumitomo[d]
(R)-N-DNB-phenylglycine (covalent)*	Covalent D-Phenyl Glycine	Regis[a]
	DNBPG (covalent)	Baker[b]
	DNBPG-C-Si100Polyol	Serva[e]
	Sumipax 0A2000A	Sumitomo[d]
(S)-N-DNB-phenylglycine (covalent)*	Covalent L-Phenyl Glycine	Regis[a]
(R)-N-DNB-leucine (covalent)*	DNBDL-C-Si100Polyol	Serva[e]
(S)-N-DNB-leucine (ionic)*	Ionic L-leucine	Regis[a]
(S)-N-DNB-leucine (covalent)*	Covalent L-leucine	Regis[a]
	Covalent DNBLeu	Baker[b]
	DNBLL-C-Si100 Polyol	Serva[e]
(R, S)-DNB-phenylglycine (covalent)*	Covalent D,L-Phenyl Glycine	Regis[a]
(R)-naphthylalanine**	L-Naphthyl Alanine	Regis[a]
(S)-naphthylalanine**	D-naphthyl Alanine	Regis[a]
(R, S)-naphthylalanine**	D,L-naphthyl Alanine	Regis[a]
(S)-1-(α-naphthyl)ethylamine**	Sumipax 0A-1000	Sumitomo[d]
	Sumipax 0A-1000A	Sumitomo[d]
N-(S)-2-(4 chlorophenyl)isovaleroyl -(R)-phenylglycine**	Sumipax 0A-2100	Sumitomo[d]

N-(1R,3R)-*trans*-chrysanthemoyl -(R)-phenylglycine	Sumipax OA-2200	Sumitomo[d]
N-*tert*-butylaminocarbonyl-(S)-valine	Sumipax OA-3000	Sumitomo[d]
N-(S)-1-(α-naphthyl)ethylaminocarbonyl -(S)-valine*	Sumipax OA-4000	Sumitomo[d]
N-(R)-1-(α-naphthyl)ethylaminocarbonyl -(R)-valine*	Sumipax OA-4100	Sumitomo[d]
(R)-(1-phenyl)ethylurea	Supelcosil LC-(R)-Urea	Supelco[f]
Cyclodextrin bonded stationary phases		
α-Cyclodextrin	Cyclobond III	Astec[g]
β-Cyclodextrin	Cyclobond I	Astec[g]
	Chiral BDex=Si100 Polyol	Serva[e]
Acetylated-β-cyclodextrin	Acetylated-Cyclobond I	Astec[g]
γ-Cyclodextrin	Cyclobond II	Astec[g]
Protein Bonded Stationary Phases		
Bovine serum albumin	Resolvosil	Macherey Nagel[h]
α-Acid glycoprotein	Enantiopac	LKB[k]
Chiral Polymer Stationary Phases		
Cellulose triacetate	Cellulose Cel-AC-40 XF**	Macherey Nagel[h]
	Hibar*RT Microcrystalline**	EM[l]
	Cellulose triacetate	
	Chiracel OA	Daicel[m]
Cellulose tribenzoate	Chiracel OB	Daicel[m]
Cellulose trisphenylcarbonate	Chiracel OC	Daicel[m]
Cellulose tribenzyl ether	Chiracel OE	Daicel[m]
Cellulose tricinnamate	Chiracel OK	Daicel[m]
Poly(triphenylmethyl methacrylate)	Chiralpak OT(+)	Daicel[m]

Table 5 Continued

Types of chiral stationary phases	Trade name	Manufacturer
Poly(2-pyridyldiphenylmethyl methacrylate)	Chiralpak OP (+)	Daicel[m]
Polyamide	Chiralspher	EM[l]
Ligand exchange stationary phases		
Proline–copper	Chiralpak WH	Daicel[m]
	Chiral ProCu=Si100Polyol	Serva[e]
Hydroxyproline–copper	Chiral HyproCu=Si100 Polyol	Serva[e]
	Nucleosil Chiral-1	Macherey Nagel[h]
Valine–copper	Chiral ValCu=Si100 Polyol	Serva[g]
Amino acid–copper	Chiralpak WM	Daicel[m]

DNB = (3,5 dinitrobenzoyl).
*π acceptor.
**π donor.
[a]Morton Grove, Illinois.
[b]Phillipsburg, New Jersey.
[c]Deerfield, Illinois.
[d]Osaka, Japan.
[e]Heidelberg, Germany.
[f]Bellefonte, Pennsylvania.
[g]Whippany, New Jersey.
[h]Duren, Germany.
[k]Gaithersburg, Maryland.
[l]Cherry Hill, New Jersey.
[m]Tokyo, Japan.

Pirkle Type Chiral Stationary Phases

A series of chiral stationary phases capable of separating a large range of enantiomers was designed by Pirkle and coworkers based on a three-point chiral recognition model proposed by Dalgliesh [53,54]. According to the model, at least three simultaneous interactions between the chiral stationary phase and the enantiomeric solute are required for chiral resolution. Furthermore, at least one of these interactions, attractive or repulsive, must be stereochemically controlled.

The two general forms of these stationary phases, which differ by whether they are ionically or covalently bonded to the silica backbone, are illustrated in Figure 2. These stationary phases contain a number of possible sites for interactions: (a) π-π bond donor-acceptor interaction from the 3,5 dinitrobenzoyl ring, (b) hydrogen bonding from the amide hydrogen and carboxyl, (c) dipole formed at the amide linkage, and (d) possible van der Waals attractive π-π interaction and/or steric repulsion (due to the relative bulkiness of the group between the R group). These sites provide a number of ways for the chiral stationary phase to interact with the enantiomeric solute. These enantiomeric solutes include a wide range of molecules [55]. A wide variety of Pirkle type stationary phases are now commercially available (see Table 5). They can be classified as either π-electron acceptor or π-electron donor stationary phases. In general the π-electron acceptor stationary phases are used to resolve π-electron donors such as aromatic alcohols, aromatic amines, aromatic sulfoxides, hydantoins, succinimides, and lactams [54], while the π-electron donor stationary phases are employed to resolve amines, alcohols, amino acids, and amino alcohols that have been derivatized with a π-electron acceptor such as 3,5 dinitrobenzoyl chloride [66,67].

The main advantage of the Pirkle type chiral stationary phases is their broad applicability in resolving a wide range of compounds, as illustrated by the different

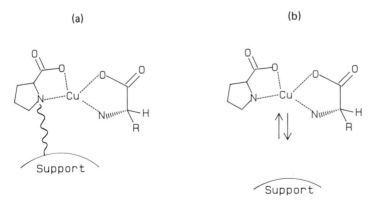

Figure 2 General structure of (a) ionically and (b) covalently bonded Pirkle type stationary phases.

classes of drugs reported in Table 6. Furthermore, relatively high chromatographic efficiency and enantiomeric selectivity have been routinely observed in applications using these stationary phases [56-64]. The major disadvantage of these chiral stationary phases is that enantiomers containing polar groups such as amino or carboxy functions usually have to be derivatized in order to obtain satisfactory enantiomeric resolution [56-64]. In addition, some of these stationary phases are not compatible with aqueous mobile phases and thus are of limited use in the analysis of drugs in biological fluids.

Recent research to improve the applicability of this type of packing has involved the further development of modified chiral stationary phases [65-69], the investigation of newer derivatizing reagents [64,70], and the evaluation of mobile phases for optimizing enantiomeric selectivity [71].

Cyclodextrin Bonded Stationary Phases

Chiral cyclodextrins are cyclic oligosaccharides consisting of six or more D-(+)-glycopyranose units. They have the shape of a hollow truncated cone with the larger end ringed with the secondary hydroxyl groups and the smaller end rimmed with primary hydroxyl groups (see Figure 3). The interior of the cavity, which contains glycosidic oxygen bridges, is, therefore, relatively hydrophobic [46]. The smallest cyclodextrin homologs, α-, β-and γ-cyclodextrins, which consist of six, seven, and eight glucose units respectively, are commercially available.

Although various cyclodextrin stationary phases have been synthesized by either cross-linking cyclodextrin polymers or chemically attaching cyclodextrins to silica via different bonding processes [72-75], only the cyclodextrin bonded stationary phases developed by Armstrong and coworkers are commercially available [75,76]. These stationary phases, which are synthesized by chemically linking cyclodextrin to silica via a hydrolitically stable and non-nitrogen-containing linkage, have found widespread applications in resolving various optical, geometrical and structural isomers as well as other routine compounds [76-79].

The basic property of cyclodextrins that allows them to affect numerous chemical separations is their ability to include selectively, in solution, a variety of guest molecules in their cavities. The formation of the inclusion complex may be caused by either a hydrophobic effect (due to favorable dipole-dipole interaction between the cyclodextrin and the guest molecule), hydrogen bonding (between the cyclodextrin's hydroxyl groups and the guest molecule), the release of high energy water or organic modifier during complex formation, and/or a combination of these factors [46]. Since inclusion complex is a spatial interaction, complexes formed between the cyclodextrin and the two forms of an enantiomer may have different stabilities. The basis for chiral recognition by cyclodextrins can be rationalized by Dalgliesh's three-point interaction concept [53]. The three points of interaction consist of a dipole-dipole interaction with the cyclodextrin's cavity and two other interactions with the cyclodextrin's rim. The two interactions with the rim can be attractive hydrogen bondings and/or repulsive steric interactions.

Table 6 Enantiomeric Resolution of Pharmaceuticals Using Pirkle Type Chiral Stationary Phases

Pharmaceuticals	Derivatizing reagent	Stationary phases[a]	References
α-Methyl arylacetic acids (ibuprofen, naproxen, fenoprofen, benoxaprofen)	Arylchlorides/1-napthalenemethylamine	2	56
Tropicamide	Direct	2	57
Tropic acid	Arylchlorides/1-napthalenemethylamine		
Amphetamine and analogs	β-Napthylchloroformate	1	64
	2-Napthylacetylchloride	1, 2	58
Propranolol	Phosgene	1	59
Benzodiazepin-2-ones	Direct	2, 3	60
Barbiturates	Direct	1–4	61
Succimimides/mephenytoin	Direct	1–4	61
a-β Amino alcohols (ephedrine, pseudoephredrine, 4-methoxylpherine, etc.)	2-Napthaldehyde	1	62
	Other aromatic aldehydes		63

[a] 1, ionically bonded (R)-N-(3, 5 dinitrobenzoyl) phenylglycine; 2, covalently bonded (R)-N-(3, 5 dinitrobenzoyl) phenylglycine; 3, covalently bonded (S)-N-(3, 5 dinitrobenzoyl) leucine; 4, ionically bonded (S)-N-(3, 5 dinitrobenzoyl) leucine.

Various enantiomeric separations of pharmaceuticals using cyclodextrin bonded stationary phases are summarized in Table 7. Most of these separations were achieved using aqueous mobile phases and the β-cyclodextrin bonded stationary phase, and without the need to do precolumn derivatization of the drugs. Future development on the applications of various cyclodextrin bonded stationary phases with different cavity sizes and functional groups will undoubtedly expand the usefulness of this type of stationary phase in pharmaceutical separations.

Protein Bonded Stationary Phases

The ability of proteins stereoselectively to bind low molecular weight enantiomers has been observed in various protein-drug equilibria studies [82]. This remarkable ability of proteins originates from their complex three-dimensional structures, which contain various functional groups capable of manifesting different types of interactions. This feature was used by two different research groups in designing their chiral stationary phases [83,84]. One type of protein column was developed by Allenmark et al. by covalently bonding bovine serum albumin to silica gel [83]. Another type of protein bonded stationary phase was synthesized by Hermansson using ionic bonding and then cross-linking α-acid glycoprotein to silica gel [84]. Both types of stationary phases have demonstrated unusual ability to resolve enantiomers. The various enantiomeric separations using the two chiral stationary phases are summarized in Table 8. Even though both of these two stationary phases are now

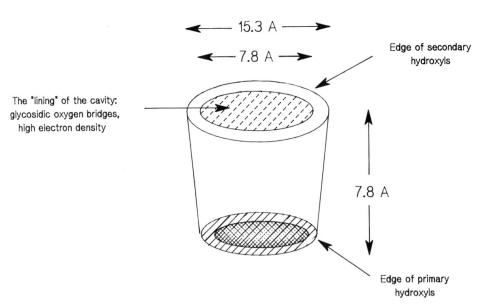

Figure 3 β-Cyclodextrin structure.

Table 7 Enantiomeric Resolution of Pharmaceuticals Using Cyclodextrin Bonded Stationary Phases

Stationary phases	Pharmaceuticals	References
β-Cyclodextrin	Propranolol, metoprolol, chlorpheniramine, verapamil, nisolidipene, nimodipene, chlorthalidone, hexobarbital, mephobarbital, mephenytoin, phensuccimide, aminoglutethimide, ketoprofen, methadone, and methylpehinidate	79
Acetylated β-cyclodextrin	Norgestrel	80

Table 8 Enantiomeric Resolution of Pharmaceuticals Using Protein Bonded Stationary Phases

Stationary phases	Pharmaceuticals	References
Bovine serum albumin	Prilocain, phenprocoumon, warfarin, and benzodiazepine derivatives	84
α-Acid glycoprotein	40 basic drugs with widely different structures	85
α-acid glycoprotein	8 acidic drugs with widely different structures	86

commercially available, to date the α-acid glycoprotein bonded stationary phase appears to offer a wider range of application to molecules of pharmacological interest. As in the case of cyclodextrin bonded stationary phases, many separations performed using these two stationary phases were achieved without the need to derivatize the enantiomeric solutes [84–87]. Furthermore, these columns allow the use of aqueous mobile phases, which are preferred for drugs in biological fluids. Also, this approach has the advantages of optimizing the solute retentions as well as the enantioselectivity by adjusting such separation parameters as ionic strength, pH, and ion pair reagents [84–87].

Chiral Polymer Stationary Phases

The chiral recognition ability of optically active polymers such as cellulose, cellulose derivatives, and polyamides has been known for some time [89,90]. However, until recently, most enantiomeric separations using chiral polymers were performed using planar or low pressure column chromatography. Part of the reason may be caused by the difficulty in preparing cross-linked polymers having sufficient mechanical strength, minimal swelling in various solvents, and appropriate size for high pressure liquid chromatography. Consequently, simple alternative approaches for preparing chiral polymer stationary phases, which involve either physically coating or covalently bonding the chiral polymers to a macroporous support, have been developed. Indeed, macroporous silica coated with poly (triphenylmethyl methacrylate) and cellulose derivatives [91–93] or covalently bonded with polyacrylamides [94] have demonstrated remarkable ability in resolving various enantiomers. Recent enantiomeric separations using various chiral polymer stationary phases are summarized in Table 9. It should be noted that the coated polymer packings usually exhibited different selectivity from their analogous packings prepared by cross linking pure polymers [93]. In the case of coated polymer stationary phases, only mobile phases consisting of non-solvents of the chiral polymer can be used to prevent its leaching [93]. The structures of several cellulose derivatives used in the preparation of commercially available stationary phases are illustrated in Figure 4.

There are controversies concerning the mechanism for chiral recognition by chiral polymers [94,95]. Nevertheless, it appears that chiral recognition does not result from the direct interactions between the individual optically active residue of the chiral polymers and the enantiomers but rather from the inclusion of the enantiomers into the asymmetric cavities of the three-dimensional polymeric network [94]. Since several chiral polymeric stationary phases are now commercially available (see Table 5), numerous applications using this type of stationary phase will undoubtedly appear in the near future.

Other Chiral Stationary Phases

Several other types of stationary phases have been known to resolve enantiomers, for example, the charge-transfer stationary phases (synthesized by bonding or coating charge-transfer adducts such as α-(2,4,5,7-tetranitro-9-fluorylideneaminooxy)

Table 9 Enantiomeric Resolution of Pharmaceuticals Using Chiral Polymer Stationary Phases

Stationary phases	Pharmaceuticals	References
Cellulose carbamate polymer	Antiepileptic and hyponotic agents (mephobarbital, ethotoin, etc.)	96
Poly(acryamide) polymer	Chlothalidone, ifosfamide	94
Poly(methacrylamide) polymer	Thalidomide and analogues	94
Microcrystalline cellulose triacetate	Barbiturates, benzothiodiazepines, retamin, mianserin, praxignantel, oxapadol, volipram, methaqualone, and chlormezanon	94
Poly(acryloyl-S-phenylalanine-ethylester) bonded to silica	Benzodiazepines	94

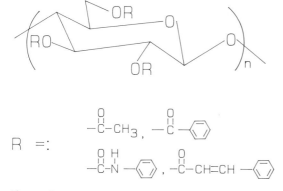

Figure 4 The structure of several cellulose derivatives used in the preparation of Daicel's Chiralcel stationary phases.

propionic acid to silica gel [97-99] and the crown ether bonded stationary phases [99-101]. These stationary phases, unfortunately, have very limited applicability and have not been used for the analysis of enantiomeric drugs.

CONCLUSION

It is clear that the range of liquid chromatographic techniques for enantiomeric analysis is now very broad. This is because no single liquid chromatographic technique can provide superior and universal chiral separation of all classes of compounds. Consequently, chiral liquid chromatographic separation is not trivial, since it is still difficult to decide which technique would work for an unfamiliar chiral compound. Undoubtedly, further advances in the development of more universal chiral liquid chromatographic techniques and in the understanding of chiral recognition mechanisms are needed before chiral liquid chromatography can be considered routine.

ACKNOWLEDGMENT

The author is grateful to Mrs. Barbara Costello for typing this manuscript.

REFERENCES

1. A. M. Barrett and V. A. Cullum, The biological properties of the optical isomers of propranolol and their effects on cardiac arrhythmias, *Br. J. Pharmacol., 34*:43 (1968).
2. M. S. Lennard, G. T. Tucker, J. H. Silas, S. Freestone, L. E. Ramsay, and H. F. Woods, Differential stereoselective of metoprolol in extensive and poor debrisoquine metabolisers, *Clin. Pharmal. Ther., 34*:732 (1983).

3. C. Von Bahr, J. Hermansson, K. Tawara, Plasma Levels of (+) and (−) propranolol and 4-hydroxypropranolol after administration of racemic (+) propranolol in man, *Brit. J. Clin. Pharmal.,* 14:79 (1982).
4. M. S. Lennard, G. T. Tucker, and H. F. Woods, The polymorphic oxidation of β_2-adrenoceptor antagonist, *Clin. Pharmacokin.,* 11:1 (1986).
5. W. Lindner, C. Leitner, and G. Uray, Liquid chromatographic separation of enantiomeric alkanoamines via diastereomeric tartaric acid monoesters, *J. Chromatogr.,* 316:605 (1984).
6. B. Halpern, Derivatives for chromatographic resolution of optically active compounds, *Handbook of Derivatives for Chromatography,* (K. Blau and G. S. King, eds.), Heyden, London, p. 457 (1977).
7. E. Gil-Av and D. Nurok, Present status of enantiomeric analysis by gas chromatography, *Adv. Chromatogr.,* 10:99 (1975).
8. J. Hermansson and C. Von Bahr, Simultaneous determination of *d*- and *l*-propranolol in human by high-performance liquid chromatography, *J. Chromatogr.,* 221:109 (1980).
9. I. S. Krull, The liquid chromatographic resolution of enantiomers, *Adv. Chromatogr.,* 16:416 (1980).
10. G. Helmchen, G. Nill, D. Flockerzi, W. Shukle, and M. Yousef, Preparative scale directed resolution of enantiomeric carboxylic acids and lactones via liquid chromatography, *Angw. Chrm. Int. Ed. Engl.,* 18:12 (1979).
11. W. H. Pirkle and J. R. Hauske, Broad spectrum methods for the resolution of optical isomers: A discussion of the reasons underlying the chromatographic separability of some diastereomeric carbamates, *J. Org. Chem.,* 42:1839 (1977).
12. A. J. Sedman and J. Gal, Resolution of the enantiomers of propranolol and other beta-adrenergic antagonists by high-performance liquid chromatography, *J. Chromatogr.,* 278:199 (1983).
13. N. Minura, Y. Kasahara, and T. Kinoshita, Resolution of the enantiomers of norepinephrine and epinephrine by reversed-phase high-performance liquid chromatography, *J. Chromatogr.,* 213:337 (1981).
14. J. Gal, Resolution of the enantiomers of ephedrine, norephedrine and pseudophredrine by reversed-phase liquid chromatography, *J. Chromatogr.,* 307:330 (1984).
15. J. Gal, Determination of the enantiomeric composition of chiral aminoalcohols using chiral derivatization and reversed phase liquid chromatography, *J. Liq. Chromatogr.,* 9:673 (1986).
16. J. Gal and A. J. Sedman, R-α-methylbenzyl isothiocyanate, a new and convenient chiral derivatizing agent for the separation of enantiomeric amino compounds by high-performance liquid chromatography, *J. Chromatogr.,* 314:275 (1984).
17. J. Thompson, J. Holtzman, M. Tsuru, C. Lerman, and J. Holtzman, Procedure for chiral derivatization and chromatographic resolution of *R* and *S* propranolol, *J. Chromatogr.,* 238:470 (1982).
18. W. Dieterle and J. W. Faigle, Multiple inverse isotope dilution assay for oxprenolol and nine metabolites in biological fluids, *J. Chromatogr.,* 259:301 (1983).
19. W. Dieterle and J. W. Faigle, Multiple inverse isotope dilution assay for the stereospecific determination of *R* (+)-and *S* (−)-oxprenolol in biological fluids, *J. Chromatogr.,* 259:311 (1983).

20. C. R. Clark and J. M. Bardsdale, Synthesis and liquid chromatographic evaluation of some chiral derivatizing agents for resolution of amine enantiomers, *Anal. Chem.*, *56*:958 (1984).
21. J. Goto, N. Goto, and T. Nambara, Sensitive derivatization reagents for the resolution of carboxylic acid enantiomers by high-performance liquid chromatography, *J. Chromatogr.*, *239*:559 (1982).
22. J. Goto, N. Goto, T. Nishimaki, and T. Nambara, Sensitive derivatization reagents for the resolution of carboxylic enantiomers by high-performance liquid chromatography, *Anal. Chim. Acta*, *120*:187 (1980).
23. J. Goto, M. Ito, S. Katsuki, N. Saito, and T. Nambara, Sensitive derivatization reagents for optical resolution of carboxylic acids by high performance liquid chromatography with fluorescence detection, *J. Liq. Chromatogr.*, *9*:683 (1986).
24. J. Goto, M. Kasegawa, S. Nakumara, K. Shimada, and T. Nambara, New Derivatization reagents for the resolution of amino acid enantiomers by high-performance liquid chromatography, *J. Chromatogr.*, *152*:413 (1978).
25. V. A. Davankov and S. V. Rogozlin, Ligand chromatography as a novel method for the investigation of mixed complexes: Stereoselective effects in α-amino acid copper(II) complexes, *J. Chromatogr.*, *60*:280 (1971).
26. V. A. Davankov, S. V. Rogozlin, A. V. Semechkin, and T. P. Sachkova, Ligand exchange liquid chromatography of racemates: Influence of the degree of saturation of the asymmetric resin by metal ions on ligand exchange, *J. Chromatogr.*, *82*:359 (1973).
27. V. A. Davankov, Ligand exchange chromatography, *Adv. Chromatogr.*, *18*:139 (1980).
28. V. A. Davankov, A. A. Kurganov and A. S. Bochkov, Resolution of racemates by high performance liquid chromatography, *Adv. Chromatogr.*, *22*:71 (1983).
29. O. Gubidz, W. Jellenz, and W. Santi, Separation of the optical isomers of amino acids by ligand exchange chromatography using chemically bonded chiral phases, *J. Chromatogr.*, *203*:377 (1981).
30. J. Bouë, R. Andebert, and G. Guivoron, Direct resolution of α-amino acid enantiomers by ligand exchange: Stereoselection mechanism on silica packings coated with a chiral polymers, *J. Chromatogr.*, *204*:185 (1981).
31. B. Feibush, M. J. Cohen, and B. L. Karger, The role of bonded phase composition on the ligand exchange chromatography of dansyl D-L amino acids, *J. Chromatogr.*, *282*:3 (1983).
32. G. Gubitz, Direct separation of enantiomers by high performance ligand exchange chromatography on chemically bonded chiral phases, *J. Liq. Chromatogr.*, *9*:519 (1986).
33. J. Lepage, W. Lindner, G. Davies, D. Seitz, and B. Karger, Resolution of the optical isomers of dansyl amino acids by reversed phase liquid chromatography with optically active metal chelate additives, *Anal. Chem.*, *51*:4533 (1979).
34. W. Lindner, J. LePage, G. Davies, D. Seitz, and B. Karger, Reversed-phase separation of optical isomers of DNS amino acids and peptides using chiral metal chelate additives, *J. Chromatogr.*, *185*:323 (1979).
35. P. E. Hare and E. Gil-Av, Separation of *D* and *L* amino acids by liquid chromatography: Use of chiral eluent, *Science*, *204*:1226 (1979).
36. W. Lindner and I. Hirschbloch, Chromatographic resolution of amino acids using tartaric acid mono-*n*-octylamide as mobile phase additives, *J. Liq. Chromatogr.*, *9*:551 (1986).

37. E. Oelrich, H. Preusch and E. Wilkelm, Separation of enantiomers by high performance liquid chromatography using chiral eluents, *J. High Resolut. Chromatogr.*, *32*:269 (1980).
38. S. Lam and A. Karmen, Resolution of optical isomers as the mixed chelate copper (II) complexes by reversed phase chromatography, *J. Liq. Chromatogr.*, *9*:291 (1986).
39. R. Horikawa, R. Sakamato, and T. Tanimura, Separation of a-hydroxy acid enantiomers by high performance liquid chromatography using copper (II) amino acid eluent, *J. Liq. Chromatogr.*, *9*:297 (1986).
40. C. Gilon, R. Leshem, Y. Tapuhi, and E. Gruskha, Reversed phase chromatographic resolution of amino acid enantiomers with metal aspartame eluents, *J. Am. Chem. Soc.*, *101*:7612 (1979).
41. C. Pettersson and G. Schill, Separation of enantiomers in ion pair chromatographic system, *J. Liq. Chromatogr.*, *9*:269 (1986).
42. C. Pettersson, Chromatographic separation of enantiomers of acids with quinine as chiral counter ion, *J. Chromatogr.*, *316*:553 (1984).
43. C. Pettersson and G. Shill, Separation of enantiomeric amines by ion-pair chromatography, *J. Chromatogr.*, *204*:179 (1981).
44. C. Pettersson and K. No, Chiral resolution of carboxylic and sulphonic acids by ion-pair chromatography, *J. Chromatogr.*, *282*:671 (1983).
45. C. Pettersson and H. W. Stunrman, Direct separation of enantiomers of ephedrine and some analogues by reversed phase liquid chromatography using (+)di-*n*-butytartrate as liquid stationary phase, *J. Chromatogr. Sci.*, *22*:441 (1984).
46. W. L. Hinze, Applications of cyclodextrin in chromatographic separation and purification methods, *Sep. Purif. Methods*, *10*:51 (1981).
47. E. M. Sellers and J. Wenr Koch, Interaction of warfarin stereoisomers with human albumin, *Pharm. Res. Comm.*, *7*:311 (1975).
48. E. Albani, R. Riva, M. Custin, and A. Baruzzi, Stereoselective binding of propranolol to human a-acid glycoprotein, *Br. Jr. Clin. Pharmacol.*, *18*:244 (1975).
49. D. Sybilska, J. Zukowski, and J. Bogarski, Resolution of mephenytoin and some chiral barbiturates into enantiomers by reversed phase high performance liquid chromatography via β-cyclodextrin inclusion complexes, *J. Liq. Chromatogr.*, *9*:591 (1986).
50. J. Debowski, D. Sybilska and J. Juraczak, β-cyclodextrin as a chiral component of the mobile phase for separation of mandelic acid into enantiomers in reversed-phase systems of high-performance liquid chromatography, *J. Chromatogr.*, *237*:303 (1982).
51. J. Debowski, D. Sybilska and J. Jurczak, Resolution of some chiral mandelic acid derivatives into enantiomers by reversed-phase liquid chromatography via α- and β-cyclodextrin inclusion complexes, *J. Chromatogr.*, *282*:83 (1983).
52. G. Szepesi, M. Gazdaz, and L. Huozar, The use of cyclodextrin as mobile phase additives for the HPLC separation of enantiomeric compounds, 10th Int. Symp. Col. Liq. Chromatogr., San Francisco, California (1986).
53. C. E. Dalgliesh, Optical resolution of amino acids on paper chromatograms, *J. Chem. Soc.*, *137*:3940 (1952).
54. W. H. Pirkle, J. M. Finn, J. L. Schreiner, and B. C. Hamper, A widely useful chiral stationary phase for the high performance liquid chromatography of enantiomers, *J. Am. Chem. Soc.*, *103*:3964 (1981).
55. I. W. Wainer and T. D. Doyle, Stereoisomeric separations, *Liq. Chromatogr.*, *2*:88 (1984).

56. I. W. Wainer and T. D. Doyle, Application of high-performance liquid chromatographic chiral stationary phases to pharmaceutical analysis: Structural and conformational effects in the direct enantiomeric resolution of α-methyl acrylacetic acid antiinflammatory agents, *J. Chromatogr.*, 284:11 (1984).
57. I. W. Wainer and T. D. Doyle, Application of HPLC chiral stationary phases to pharmaceutical analysis: The resolution of some tropic acid derivatives, *J. Liq. Chromatogr.*, 7:731 (1984).
58. I. W. Wainer and T. D. Doyle, Application of high-performance liquid chromatographic stationary phases to pharmaceutical analysis: The direct enantiomeric resolution of amide derivatives of 1-phenyl-2-aminopropane, *J. Chromatogr.*, 259:465 (1983).
59. I. W. Wainer, T. D. Doyle, K. H. Down, and J. R. Powell, The direct enantiomeric determination of (-) and (+) propranolol in human serum by high-performance liquid chromatography on a chiral stationary phase, *J. Chromatogr.*, 306:405 (1984).
60. W. H. Pirkle and A. Tsipouras, Direct liquid chromatographic separation of benzodiazepinone enantiomers, *J. Chromatogr.*, 291:291 (1984).
61. Z. Y. Yang, S. Barkan, C. Brunner, J. D. Weber, T. D. Doyle, and I. W. Wainer, Application of high-performance liquid chromatographic stationary phases to pharmaceutical analysis: Resolution of enantiomeric barbiturates, succimides and related molecules on four commercially available chiral stationary phases, *J. Chromatogr.*, 324:444 (1985).
62. I. W. Wainer, T. D. Doyle, Z. Hamidzadeh, and M. Alridge, Application of high performance liquid chromatographic chiral stationary phases to pharmaceutical analysis: Resolution of ephedrine, *J. Chromatogr.*, 261:123 (1983).
63. I. W. Wainer, T. D. Doyle, F. S. Fry, and Z. Hamidzadeh, Chiral recognition model for the resolution of ephedrine and related α, β aminoalcohols as enantiomeric oxazolidine derivatives, *J. Chromatogr.*, 355:149 (1986).
64. T. D. Doyle, W. M. Adams, F. S. Fry, and I. W. Wainer, The application of HPLC chiral stationary phases to stereochemical problems of pharmaceutical interest: A general method for the resolution of enantiomeric amines and β-napthyl carbamate derivatives, *J. Liq. Chromatogr.*, 9:455 (1986).
65. W. H. Pirkle, M. H. Hyun, and B. Bank, A rational approach to the design of highly effective chiral stationary phases, *J. Chromatogr.*, 316:585 (1984).
66. N. Oi and H. Hitakara, Enantiomer separation by HPLC with some urea derivatives of L-valine as novel chiral stationary phases, *J. Liq. Chromatogr.*, 9:511 (1986).
67. W. H. Pirkle and T. C. Pochapsky, A new, easily accessible reciprocal chiral stationary phase for the chromatographic separation of enantiomers, *J. Am. Chem. Soc.*, 108:352 (1986).
68. N. Oi and H. Hitakara, High-performance liquid chromatographic separation of amino acid enantiomers on urea derivatives of L-valine bonded to silica gel, *J. Chromatogr.*, 285:198 (1984).
69. R. Dappen, V. R. Mayer, and H. Arm, Chiral covalently bonded stationary phases for the separation of enantiomeric amine derivatives by high-performance liquid chromatography, *J. Chromatogr.*, 295:367 (1984).
70. W. H. Pirkle, G. Mahler, and M. H. Hyun, Separation of the enantiomers of 3,5-dinitrophenyl carbamates and 3,5-dinitrophenyl ureas, *J. Liq. Chromatogr.*, 9:443 (1986).
71. M. Zief, L. J. Crane, and J. Horvath, Selection of mobile phase for the enantiomeric resolution via chiral stationary phase columns, *J. Liq. Chromatogr.*, 7:709 (1984).

72. B. Zsadon, L. Decsel, M. Szilaris, F. Tudos, and J. Szegtli, Inclusion chromatography of enantiomers of indole alkaloids on cyclodextrin polymer stationary phase, *J. Chromatogr.,* 270:127 (1983).
73. K. Fujimura, T. Meda, and T. Ando, Retention behavior of some aromatic compounds on chemically bonded cyclodextrin silica stationary phases in liquid chromatography, *Anal. Chem.,* 55:446 (1983).
74. Y. Kawaguchi, M. Tanaka, M. Nakal, K. Funazo, and T. Shono, Chemically bonded cyclodextrin stationary phases for liquid chromatographic separation of aromatic compounds, *Anal. Chem.,* 55:1852 (1983).
75. D. W. Armstrong, U.S. Patent No. 4539339 (1985).
76. D. W. Armstrong and W. Demond, Cyclodextrin bonded phases for the liquid chromatographic separation of optical, geometrical and structural isomers, *J. Chromatogr. Sci.,* 22:441 (1984).
77. D. W. Armstrong, W. Demond, A. Alak, W. L. Hinze, T. Riehl, and K. H. Bui, Liquid chromatographic separation of diastereomers and structural isomers on cyclodextrin bonded phases, *Anal. Chem.,* 57:234 (1985).
78. D. W. Armstrong, A. Alak, K. H. Bui, W. Demond, T. J. Ward, T. Riehl, and W. L. Hinze, Facile separation of enantiomers, geometrical isomers and routine compounds on stable cyclodextrin LC bonded phases, *J. Incl. Phenom.,* 2:533 (1984).
79. W. L. Hinze, T. E. Riehl, D. W. Armstrong, W. Demond, A. Alak, and T. J. Ward, Liquid chromatographic separation of enantiomers using a chiral β-cyclodextrin-bonded stationary phase and conventional aqueous-organic mobile phases, *Anal. Chem.,* 57:237 (1985).
80. D. W. Armstrong, T. J. Ward, R. D. Armstrong, and T. E. Beesley, Separation of drug stereoisomers by formation of β-cyclodextrin inclusion complexes, *Science,* 232:1132 (1986).
81. T. J. Ward and D. W. Armstrong, Improved cyclodextrin chiral phases: A comparison and review, *J. Liq. Chromatogr.,* 9:407 (1986).
82. C. F. Chignell, Drug-protein interactions, *Handbook of Biochemistry and Molecular Biology: Proteins* (G. D. Fasman, ed.) CRC Press, Cleveland, Ohio, p. 544 (1976).
83. S. Allenmark, B. Bomgren, and H. Boren, Direct resolution of enantiomers by liquid affinity chromatography on albumin agarose as the chiral stationary phase, *J. Chromatogr.,* 237:473 (1982).
84. J. Hermansson, Direct liquid chromatographic resolution of racemic drugs using α2-acid glycoprotein as the chiral stationary phase, *J. Chromatogr.,* 269:71 (1983).
85. S. Allenmark, Optical resolution by liquid chromatography on immobilized bovine serum albumin, *J. Liq. Chromatogr.,* 9:425 (1986).
86. G. Schill, I. W. Wainer, and S. A. Barban, Direct liquid chromatographic resolution of cationic drugs using a chiral α-acid glycoprotein bonded stationary phase, *J. Liq. Chromatogr.,* 9:641 (1986).
87. J. Hermansson and M. Erikson, Direct liquid chromatographic resolution of acidic drugs using a chiral α-acid glycoprotein column, *J. Liq. Chromatogr.,* 9:621 (1986).
88. M. Katake, T. Sakan, N. Nakamura, and S. Senon, Resolution into optical isomers of amino acids by paper chromatography, *J. Am. Chem. Soc.,* 73:2973 (1951).
89. K. Bach and J. Haas, Dünnschichtchromatographische Spaltung der Racemate einiger Aminosäuren, *J. Chromatogr.,* 136:186 (1977).
90. G. Blaschke, New analytical method. Chromatographic racemate resolution, *Angw. Chem. Int. Ed. Engl.,* 19:13 (1980).

91. Y. Okamoto, I. Okamoto, H. Yuki, S. Murata, R. Noyor, and H. Takaya, New novel packing material for optical resolution: (+) poly(triphenylmethyl) methacrylate coated on macroporous silica gel, *J. Am. Chem. Soc., 103*:6971 (1981).
92. Y. Okamoto, E. Yashima, K. Halada, and K. Mislow, Chromatographic resolution of perchloro triphenylamine on (+) poly(triphenylmethyl) methacrylate, *J. Org. Chem., 49*:557 (1984).
93. Y. Okamoto and K. Halada, Resolution of enantiomers by HPLC on optically active poly(triphenylmethyl) methacrylate, *J. Liq. Chromatogr., 9*:341 (1986).
94. G. Blaschke, Chromatographic resolution of chiral drugs on polyamides and cellulose triacetate, *J. Liq. Chromatogr., 9*:341 (1986).
95. T. Shibata, I. Okamoto, and K. Ishu, Chromatographic optical resolution on polysaccharides and their derivatives, *J. Liq. Chromatogr., 9*:313 (1986).
96. *Chiral Pak and Chiral Cel Technical Brochure No. 1*, Daicel Chemical Industries, LTD, Tokyo, Japan, p. 11 (1985).
97. F. Mikes and G. Boshart, Resolution of optical isomers by high-performance liquid chromatography—A comparison of two selector systems, *J. Chromatogr., 149*:455 (1978).
98. F. Mikes, G. Boshart, and E. Gil-Av, Resolution of optical isomers by high-performance liquid chromatography using coated and bonded chiral charge transfer complexing agents as stationary phases, *J. Chromatogr., 122*:205 (1976).
99. D. W. Armstrong, Chiral stationary phases for high-performance liquid chromatography of enantiomers, *J. Liq. Chromatogr., 7*:353 (1984).
100. L. R. Sousa, G. D. Y. Sogah, D. H. Hoffman, and D. J. Cram, Host-guest complexation. 12. Total optical resolution of amine and amino ester salt by chromatography, *J. Am. Chem. Soc., 100*:4569 (1978).
101. G. D. Y. Sogah and D. J. Cram, Host-guest complexation. 14. Host covalently bound to polystyrene resin for chromatographic resolution of enantiomers of amino acids and ester salts, *J. Am. Chem. Soc., 101*:3035 (1979).

11
HPLC of Proteins and Peptides in the Pharmaceutical Industry

Kalman Benedek* and Joel K. Swadesh†

Smith Kline & French Laboratories, King of Prussia, Pennsylvania

INTRODUCTION

Remarkable advances in molecular biology and biochemistry have sparked equally remarkable developments in the field of separations science. The use of cell culture for the production of genetically engineered proteins has made it possible to explore and directly intervene in the biochemical pathways that modulate disease. In recent years, we have seen the clinical successes of genetically engineered vaccines [1-5], anticancer agents [6,7], hormones [8], and thrombolytics [9,10].

Some of the major biotechnology products and some of the companies involved in the production of these compounds are listed in Table 1. More than just genetic engineering and cellular expression of the proteins of interest has been responsible for these successes. Once a protein is expressed in cell culture, purification and characterization are required before the protein can be used clinically. Analytical and preparative purification processes currently account for a large proportion of the effort and expense of the commercial production of proteins.

Separation science has played a central role in the development of the biotechnology industry. The theory and practice of high performance liquid chromatography (HPLC) techniques are now as ubiquitous in recombinant technology as in conventional "small molecule" pharmaceutics. This popularity is due to the wide applicability of HPLC as a superior analytical and preparative separation tool. HPLC

Current affliations:
*Terrapin Technologies, Inc., San Francisco, California
†University of Massachusetts at Amherst, Amherst, Massachusetts

Table 1 Some of the Products of the Biopharmaceutical Industry

Substance	Name	Type	Function	Producers
ANF	Atrial natriuretic factor	Peptide	Diuretic, anti-hypertensive	California Biotechnology
EGF	Epidermal growth factor	Protein	Wound healing	Chiron
EPO	Erythropoietin	Peptide	Stimulate red blood cell production	Amgen
Factor VIII	von Willebrand factor	Protein	Regulation of coagulation cascade	Genentech, Genetics Institute
FGF	Fibroblast growth factor	Protein	Angiogenesis	California Biotechnology
G-CSF	Granulocyte colony stimulating factor	Peptide	White blood cell stimulation, cancer therapy	Amgen
GM-CSF	Granulocyte–monocyte colony stimulating factor	Protein	White blood cell stimulator; cancer, radiation, and AIDS therapy	Genetics Institute

hGH	Human growth hormone	Peptide	Pituitary dwarfism	Genentech, Eli Lilly
IF-α, β, γ	Interferon (α, β, and γ)	Protein	Immunostimulant	Schering-Plough
IL-2	Interleukin-2	Protein	Lymphokine, chemotherapy	Cetus
IL-3	Interleukin-3	Protein	Blood cell growth	Genetics Institute
	Malaria antigens	Protein	Vaccines	SK&F, Hoffman-LaRoche, Chiron
M-CSF	Macrophage colony stimulating factor	Protein	Monocyte/macrophage stimulation	Cetus, Genetics Institute
SOD	Superoxide dismutase	Protein	Scavenges superoxide	Chiron, Bio-Technology General
TNF	Tumor necrosis factor	Protein	Lymphokine, cancer therapy	
t-PA	Tissue plasminogen activator	Protein	Fibrinolytic	Genentech, Genetics Institute

gives fast separations with high resolution and sensitivity. High performance separations are requisite to isolate a protein of interest from the complex matrix of the cell culture process, and they are involved in verifying the identity and purity of the product as well as in monitoring its stability. Due to the large variety of applicable detection methods, the utilization of HPLC techniques is almost limitless. HPLC includes such mechanistically diverse chromatographies as hydrophobic interaction chromatography (HIC), reversed phase liquid chromatography (RPLC), ion exchange chromatography (IEC), gel filtration/size exclusion chromatography (SEC), and affinity chromatography.

Prior to the use of genetic engineering to produce proteins as drug substances, a typical pharmaceutical compound might be of molecular weight 100-1500 daltons. The molecular weight of a single amino acid, the basic building block of proteins and peptides, is about 120 daltons. Therefore, a simple protein, composed of about 50-1500 amino acids, may have a molecular weight of up to about 200 kilodaltons (kD). The active pharmacological agent may be of even higher molecular weight. An example of the use of a macromolecular aggregate as a pharmaceutical is hepatitis B vaccine [11-13].

There are several chromatographic implications of the use of complex, high molecular weight compounds as pharmaceuticals. These implications are presented here in outline, and developed in subsequent sections. First, the solution properties of proteins are markedly different from those of classical drug substances. Because the molecular weights of proteins are greater than those of conventional pharmaceutical compounds, the molar solubilities of polypeptide drug substances in aqueous media tend to be lower than those of conventional compounds. Typical sample handling steps, such as dissolution, solvent exchange, and solvent removal are often accompanied by the formation of aggregates. Also, the diffusion coefficients of proteins are much smaller than those of classical drug substances, making mass transfer effects very important in chromatography. Second, polypeptide drug substances often have a high degree of heterogeneity, some of which is generated intracellularly and some of which occurs during sample handling and storage. There may be not only heterogeneity in the primary sequence as determined by the DNA but also heterogeneity generated by posttranslational hydrolysis of the backbone or modification of some of the side chain moieties, heterogeneity in the disulfide crosslinking, and heterogeneity in the attachment of prosthetic groups. The existence of heterogeneity may require that resolution of active species be suppressed even as the resolution of inactive species from active species is optimized. Third, as a direct consequence of the large conformational space permited flexible species of high molecular weights, a large polypeptide has the potential to adopt a unique three-dimensional solution structure, the adoption of which may be crucial to its function in vivo. Chromatographic handling can affect preparative yields of the active compound and analytical interpretations of the purity of the resulting isolate. Finally, there may be interactions between a polypeptide chain and its environment. In particular, further assembly of folded polypeptide chains into multisubunit species with properties different from the constituents is a common motif in protein biochemistry. Interactions of a folded polypeptide chain with the solid phase of a packaging material or of a chromatographic stationary phase can cause reversible or irreversible dissociation and unfolding. Similarly, membrane components may be associated with membrane proteins so that dissociation of one from the other may induce denaturation of the protein.

The technological requirements for the bulk production of biopharmaceuticals are far greater than those with which the pharmaceutical industry was previously faced. There are requirements in the production, analysis, and storage of peptides and proteins that are only rarely encountered in the development of conventional pharmaceutical agents. A broad understanding of the production process, from cell culture to formulation, is helpful in meeting the challenge of handling biopharmaceutical products. An intimate knowledge of the biochemistry of cellular production of proteins is helpful in identifying variants of the desired protein. An understanding of the ingredients of cell culture media and the process of harvesting helps one to anticipate the impurities that may need to be identified and removed. The development of appropriate chromatographic methodology is much easier if the production history of the biopharmaceutical, and the chemical and physical pathways of degradation and denaturation, are known. To address these issues, the first section of this chapter briefly covers basic biochemistry, including cell culture, proteolytic enzymes in cell culture, methods of protein harvesting, conformational denaturation, disulfide bond isomerism, common mechanisms of chemical degradation in processing, and sources of microheterogeneity. The second section of this chapter reviews basic protein chromatography, with attention to the solution of practical problems that occur during the production and analysis of proteins. We discuss stationary phases and how to select the right one for the desired separation; adsorption phenomena and protein unfolding; the effect of disulfide bonds on chromatographic behavior; and denaturation and degradation during chromatography. The final section examines the practical aspects of chromatographic techniques in the pharmaceutical industry.

PEPTIDES, POLYPEPTIDES, AND PROTEINS

Size, Structure, and Composition

Peptides and polypeptides are polymers of α-L-amino acids. Synthetic peptides can be linear, cyclic, or branched polymers, but the present discussion is limited to linear polymers, differing from one another in the number, type and composition of the monomers (amino acids) involved in their construction. The physicochemical properties of small peptides differ markedly from those of proteins, so it is useful to distinguish between small peptides, large peptides, and proteins. In the present article we will use the term *peptide* for biopolymers with at most 10-20 residues, and *polypeptide* for those with more than 20 amino acids. We refer to polypeptides with a unique three-dimensional structure, manifested in a definable function, as *proteins*. These are, of course, arbitrary categories.

Conformation of Biomolecules

Proteins are flexible, but they fold into specific structures. A protein typically has a characteristic, ordered, three-dimensional shape referred to as its conformation. The conformation of the protein has a vital role in areas such as catalytic activity and molecular recognition. Perturbation of the native conformation may result in irreversible conformational change and the concomitant losses of activity and specificity of substrate binding. For example, residues 57, 102, and 195 of bovine chymotrypsin I [14-16], a typical serine protease, are widely separated in the sequence but are brought into proximity by folding to form a proton transfer system. Near the catalytic site is a binding pocket that recognizes and binds aromatic amino acids. In denaturing solvents, such as the mixed aqueous-organic solvents commonly used in reversed phase liquid chromatography, the enzyme can unfold, resulting in a loss of activity.

The order of the amino acids composing the covalent backbone of the polypeptide chain is termed the primary structure of a protein. Even amino acid residues in di-and tripeptides have sterically determined conformational preferences [17]. These preferences, however, allow considerable conformational mobility, so an amino acid residue in a short peptide is, in general, very flexible. Intramolecular hydrogen bonds can typically form only in peptides with four or more residues, and the formation of regular structure occurs only in larger peptides. The most common of these regular structures are the α-helix and the β-sheet. Helices, β-sheets, and turns in the primary structure as a result of folding are secondary structural features. The complicated three-dimensional organization of the secondary structural level of an active protein is called the tertiary structure. In some cases, the in vivo assembly of multiple, independent polypeptide chains is essential for protein activity. In other cases, intermolecular association leads to inactivation and precipitation. This latter phenomenon is often referred to as aggregation, while the former is referred to as quaternary structure.

To date, the majority of proteins that have been most thoroughly described have tended to be globular proteins whose natural biological environment is aqueous, but the principles that govern protein behavior can be generalized to describe membrane-bound and fibrous proteins. As will be discussed below, the driving force of protein folding is the competition between solute-solvent and intramolecular solute-solute interactions. The formation of hydrogen bonds between the proton of a backbone amide nitrogen and the backbone carbonyl of another residue is a typical intramolecular solute-solute interaction. Such intramolecular interactions define the elements of secondary structure. Solvophobic interactions may stabilize the secondary structure, and serve to segregate, in a three-dimensional sense, solvophobic and solvophilic regions. One face of an α-helix, for example, may contain many more hydrophobic residues than the other face [18]. The interplay of solvophobic and solvophilic interactions promotes the association of specific elements of the secondary structure with one another to form the long-range order of tertiary structure. A typical motif is the association of the hydrophobic face of an α-helix with a β-sheet [19]. Another factor that may be especially important in stabilizing secondary structure,

HPLC of Proteins and Peptides

particularly the nascent secondary structure of oxidative refolding, is the interaction of disulfides and aromatic rings [20-23]. This interaction has been observed to extend chromatographic retention [24]. In the discussion below, an examination of the forces that direct protein structure is further developed.

Composition of Biomolecules

In addition to the amino acid constituents, polypeptides may be covalently or noncovalently modified by the presence of prosthetic groups. Proteins are divided into two major classes on the basis of their composition: simple and conjugated. Simple proteins are linear polymers of the common amino acids. Conjugated proteins have other organic and/or inorganic components attached through the side chain of particular amino acid residues. The attached moieties are called prosthetic groups. Conjugated proteins can be classified on the basic of the chemical nature of their prosthetic groups (e.g., nucleoproteins, lipoproteins, phosphoproteins, metalloproteins, and glycoproteins; see Table 2). Examples of such prosthetic modifications are phosphorylation of serine and threonine, glycosylation of asparagine, serine, and threonine, the attachment of heme, the binding of metals and other cofactors, and the covalent attachment of lipids. These prosthetic groups and cofactors can be important determinants in all aspects of polypeptide behavior, including secretion, activity, metabolism, and chromatographic mobility.

Glycosylation is such a common form of microheterogeneity as to warrant special attention. There is typically diversity in the carbohydrate portion of a glycoprotein. A review of asparagine-linked oligosaccharides [25] describes the biochemistry of N-linked glycosylation.

Table 2 Conjugated Proteins, Classes of Proteins with Prosthetic Group or Cofactor

Class	Examples	Prosthetic group or cofactor
Glycoprotein	γ-globulin, t-PA	Carbohydrate
Hemoprotein	Hemoglobin, myoglobin, cytochrome c, catalase	Iron protoporphyrin
Lipoprotein	B1-lipoprotein	Phospholipid, cholesterol
Metalloprotein	Ferritin	$Fe(OH)_3$
	Cytochrome oxidase	Fe, Cu
	α-lactalbumin	Ca
	Alcohol dehydrogenase	Zn
Nucleoprotein	Ribosomes	RNA
	Tobacco Mosaic virus	RNA
Phosphoprotein	Casein	Phosphate ester

Review of Cell Biology and Cell Culture

Cell Types

Two basic cell types exist, eukaryotes and prokaryotes. These are differentiated by the presence of a nuclear membrane in the eukaryotes. Prokaryotes are a primitive cell type. A typical member of this class of organisms is *Escherichia coli*. By contrast, the eukaryotes are highly diverse, functionally highly differentiated, and highly evolved. Mammalian cells are typical eukaryotes. Both cell types are widely used in cell culture techniques, and each has its advantages. Their major characteristics are listed in Table 3.

Prokaryotic cells have only one chromosome, making genetic manipulation for the production of recombinant proteins relatively easy from a technical standpoint. Prokaryotes, however, may not produce the N-terminal sequence of amino acids necessary in the eukaryotic host for the export of the polypeptide from the cell. The proteins that they produce have the correct linear sequence of amino acids, but the disulfide bonds may be in the reduced form, and the proteins are not glycosylated. Proteins are often stored as granules inside the cell, rather than as soluble products. The recovery of a product from a prokaryotic cell culture may require the disruption of the cells, releasing not only an aggregate of the product of interest, but a complex assortment of cellular debris, including DNA and proteolytic enzymes. Detergents and other solubilizing agents, such as urea and dithiothreitol, may be necessary to bring the product aggregate into solution. For complex proteins, refolding occurs in high yield only under stringently limited circumstances, and refolding is required to make an active protein. Protein variants lacking certain prosthetic moieties may be less soluble than, less active than, or metabolized differently from variants contain-

Table 3 Cell Organization in Prokaryotic and Eukaryotic Cells

	Prokaryotic cells	Eukaryotic cells
Nuclear envelope	Absent	Present
DNA	Combined with proteins	Naked
Chromosomes	Single	Multiple
Nucleolus	Absent	Present
Cell division	Amitosis	Mitosis or meiosis
Ribosomes	70S (50S + 30S)[a]	80S (60S + 40S)[a]
Membrane organelles	Absent	Present
Mitochondria	Respiratory and photosynthetic enzymes in plasma membrane	Present
Chloroplast	Absent	Present in plant cells
Locomotion	Single fibril, flagellum	Cilia and flagella

[a]S indicates Svedberg unit.

HPLC of Proteins and Peptides

ing those moieties. Some progress has been made in promoting bacterial secretion of recombinant proteins [26].

Eukaryotic cells also have disadvantages in cell culture. They are difficult to culture, many varieties requiring the presence of hormones and growth factors for proliferation [27,28]. Serum is an excellent but expensive vehicle for mammalian cell culture, with the additional disadvantage that it is a vector for a variety of serious diseases. Also, serum is an excellent medium for the growth of bacteria and fungi that spoil cell cultures.

Both prokaryotes and eukaryotes can be expected to find uses as production organisms, with the prokaryotes being especially suited to the production of simple, low molecular weight polypeptides, while eukaryotes may be best suited for the production of more elaborate protein products. The issues involved in using prokaryotes to produce eukaryotic proteins have been reviewed [29,30].

Cell Culture

Proteins are typically produced by cell culture. The matrix from which the protein of interest must be isolated includes the cell culture medium. In addition, if the protein of interest is not secreted by the cells that produce it, those cells may need to be lysed, releasing a wide variety of intracellular materials, including DNA and intracellular proteases. Proteases, either those originating in the cell culture medium or those originating intracellularly, may increase heterogeneity in the population of the protein product. Ideally, a fermentation—and mammalian cell culture should be viewed as fermentation—should be controlled by chemostasis [31] to prevent the growth of adventitious organisms. Recently, significant advances have been made in the solution of certain technical problems (such as obtaining full oxygenation of and nutrient transport to cultured cells), in the understanding of chemostatic principles, and in the acceptance of the use of cell types that require minimal media [32]. As the composition of the culture medium is simplified, so is the task of the production chemist and the analyst. Common protein additives to the culture medium are insulin, transferrin, epidermal growth factor, fibroblast growth factor, and fibronectin.

Methods of Protein Harvesting

If the protein of interest is secreted into the cell media, harvesting can be accomplished by filtration of the cells from the media. Otherwise, cell lysis may be required. Lysis can be accomplished by sonication, osmotic shock, shear, explosive decompression, or other means [33]. Lysis, of course, releases the total cellular contents, including proteases, lipids, and DNA. The release of DNA and lipids may complicate later purification, and proteases may subject the desired product to unwanted proteolysis.

Protein Refolding

Refolding In Vitro and In Vivo

Although it is possible that the pathways of in vitro refolding differ from those in vivo, the intrinsic intramolecular forces promoting folding are, presumably, unchanged. In vivo, however, it is possible that the cell is capable of compartmentalizing the refolding. One hypothesis [34] is that disulfide crosslinked proteins are, in general, conformationally too rigid to translocate across the membrane of the endoplasmic reticulum, and that disulfide-bond formation occurs posttranslocationally and therefore posttranslationally.

If a protein is produced in a foreign host, the producer organism may generate species without the proper prosthetic groups; the absence of such prosthetic groups may promote denaturation and aggregation. It is desirable to use prokaryotes as producers, because they would be unlikely to harbor viruses that would be pathogenic to man. On the other hand, prokaryotes lack the cellular machinery to handle the complex processes of selective glycosylation and oxidative refolding that eukaryotes, such as in man, require for the efficient operation of their relatively sophisticated biochemical systems.

Reversibility of Denaturation

In vitro refolding is industrially relevant and therefore of interest to the industrial chromatographer. It is not always possible or desirable to express the desired protein in an organism in which refolding occurs spontaneously, so the product may need to be artificially refolded. Also, during handling, particularly on exposure to reducing agents, to heat, to extremes of pH, or to chemical denaturants such as guanidine salts and hydrophobic chromatographic solvents, denaturation can occur. The experimentalist must be alert to the causes of, and remedies for, denaturation. Unless handling has resulted in chemical modification of the protein, the reversibility of denaturation is usually determined by the cleverness of the experimentalist in accomplishing the refolding. Heterophasic refolding, in which the reduced protein is bound to a solid support, has shown great promise as a practical approach [35]. Even in cases in which a zymogen precursor is proteolytically processed to an active form, substantial recoveries of activity by oxidative refolding have been achieved by such clever approaches as immobilization [36]. Immobilization serves to isolate individual molecules from one another to prevent aggregation.

Concepts in Refolding: Domains and Subdomains, Buried and Exposed Regions

If the size of a polypeptide is sufficiently large, different portions of a (globular) protein may exhibit regions of well-defined structure. These regions are separated from one another by amino acid residues that are not clearly incorporated into one region or another and therefore tend to be far more conformationally flexible than the residues in the structured regions. The regions of separated structure, called subdomains, can associate intramolecularly with one another to form an even more highly elaborated structure called a domain [37,38]. Only for polypeptides of degree of polymerization greater than about 200 are multiple independent domains typically observed. In this context, it should be noted that disulfide crosslinks often appear to stabilize intra-and intersubdomain interactions, and disulfide bonds are structural features common to most large proteins. At all degrees of polymerization, intermolecular association competes with intramolecular association. The most hydrophobic portions of the sequence are typically buried between regions of secondary structure. On the surface of the protein, definable regions of hydrophilic structure, called omega loops, may be seen [39].

Physical Forces in Refolding

The field of protein folding is vast and therefore almost impossible to review to detail. The principal themes of the refolding field simplify to these questions: What are the driving forces in refolding? Is there a single refolding pathway, or are multiple (or "dead-end") pathways possible? Is the native structure recovered as a highly cooperative (all-or-none) process, or do quasistable intermediates of refolding exist? What is the role of the disulfide bonds in stabilizing the native structure? Given the enormous number of structural isomers that can be generated from a sequence containing multiple cysteine residues, how is the native structure formed in high yield?

The primary sequence of a newly synthesized protein determines the tertiary structure that it will eventually adopt after refolding [40]. It seems likely that certain peptide sequences have an intrinsic tendency to form regions of ordered structure, a supposition that led to simple schemes for structural prediction [41]. By definition, however, intermediate-range interactions are necessary for the formation of structures such as β-sheets [42]. That observation, and the observation that regions of secondary structure tend to be in proximity, led to the hypothesis that short-range forces exert control over the early stages of protein folding, and that sequence-distant long-range forces dominate in the later stages of folding [43,44].

Studies of peptides derived from the N-terminal region of bovine pancreatic RNase A [45,46] indicated the presence of residual, native-like structure. The residual structure was much weaker than that observed in the native protein, indicating that intermediate-to long-range interactions help to stabilize the native structure; these observations have been rediscovered by many others [47–49]. Other regions of disulfide-reduced RNase A also probably exhibit structure, as has been observed in a study of a presumed chain folding initiation site of RNAse [50]. The complexity of the aromatic proton NMR spectrum of disulfide-reduced RNase A, and the spectral simplification that occurs on the addition of $LiClO_4$, may indicate the existence of other chain folding initiation sites.

The principal driving forces for protein refolding are the tendency to form regular hydrogen-bonded structures [51] and the tendency for the hydrophobic residues to partition inside of a globular protein [52–54]. As the hydrophobic residues are partitioned into the interior, the hydrophilic residues stabilize the partitioning by forming a water-compatible interface [55]. There is a tendency for charge pairing, particularly within α-helices [56], presumably to minimize the electrostatic energy. To account for the structure of proteins resident in membranes, the hydrophobic/hydrophilic partitioning hypothesis should, perhaps, be expanded to be solvophobic/solvophilic partitioning. One approach to the quantitation of the tendency of polypeptide sequences to form structured regions is the measurement of the Gibbs free energy of disulfide cyclization of hexapeptides in which the first and the final residues are cysteine [57].

Pathways of Refolding: Disulfide Bonds and Proline Isomers

That multiple refolding pathways exist has been experimentally confirmed by disulfide reduction and reoxidation of bovine pancreatic trypsin inhibitor (BPTI) [58] and RNase A [59], although it is also clear that oxidative refolding is a far from random process. In BPTI, the refolding process has been proven to be well directed by a number of means [60], and it is similarly evident that the process is much less well directed in RNase A [61]. One hypothesis [22] is that the tendency of disulfide bonds to cluster with aromatic residues [20,21] drives the selective formation of specific disulfide bonds.

The existence of multiple refolding pathways in disulfide-intact RNase, which may be due to proline isomerization, has also been confirmed [61,62], despite the inherent difficulty in characterizing the sparsely populated conformers generated during the refolding of the disulfide-intact protein. Refolding may be regarded as a Markovian process, governed by thermodynamics if the time scale of refolding is sufficient for equilibration between states, and governed by kinetics if it is not sufficient [63].

Sources of Heterogeneity

Protease Activity

Proteins are often produced with a signal sequence necessary for transport and are processed intracellularly to the mature protein. It has been suggested that factors governing signal sequence recognition may include the presence of a β turn [64]. When the protein is produced in a foreign host, unwanted processing can occur. In *E. coli*, the leader peptidase has been isolated from the inner membrane, sequenced, and cloned [65,66]. It is anchored by its hydrophobic *N*-terminus to the inner membrane. Numerous other proteolytic activities have been isolated from *E. coli*, including two metalloenzymes and six serine proteases [67–69]. Intracellular proteases have been recently reviewed [70].

There may be heterogeneity due to variable posttranslational cleavage of the primary sequence. In human type tissue plasminogen activator (t-PA), for example, variable cleavage has been found in the signal sequence and at Arg 276. Cleavage at residue -3 in the signal sequence generates the L (long chain) variant [71,72], and cleavage at Arg 276 generates the two-chain variant [73]. Posttranslational cleavage is not the only processing of the peptide chain that can occur. After ribosomal synthesis is complete, protein maturation can occur by protein synthesis [74,75].

Types of Heterogeneity

Heterogeneity in a class of protein species that exhibit similar activities can be divided into several types. There may be heterogeneity in the primary sequence. In principle, a cell culture could express variants containing insertions, deletions, or alterations of the sequence, and this has, in fact, been observed under certain culture conditions [76–78].

HPLC of Proteins and Peptides

There may be microheterogeneity, in which the primary sequence (as determined by the DNA sequence), is nonheterogeneous, but posttranslational or cotranslational modifications have been introduced into the side-chain moieties. Some forms of microheterogeneity, such as deamidation, are usually detectable by techniques such as isoelectric focusing, ion exchange, and chromatofocusing, whose mechanisms of separation are based on differences in charge. Such techniques may also resolve carbohydrate microheterogeneity due to the presence of sialic acid moieties of glycoproteins. Because differences in sialylation may or may not affect therapeutic performance, separations based on anionic content may need to be suppressed during production and subsequent analysis in order to maximize preparative yields.

Disulfide Bond Structural Isomers

In disulfide-containing proteins, structural isomers containing different disulfide crosslinks can exist. In small proteins, such as BPTI, the hydrodynamic volumes of disulfide structural isomers may differ significantly from that of the native protein, making SEC a potential means of separation, but SEC is not generally useful in the separation of structural isomers of significantly larger proteins. However, it is likely that the isoelectric points of disulfide isomers differ from one another, making electrostatic interaction chromatographies (EIC), ion exchange chromatography (IEC), isoelectric focusing (IEF), and chromatofocusing potential methods for separation. At present, the existence of disulfide isomers is inferred from chromatography of the fragments arising from proteolytic or chemical cleavage (peptide mapping), a procedure that is complicated by the potential for disulfide reshuffling during digestion.

Chemical Sources of Heterogeneity

During lysis and subsequent handling, the protein of interest may be exposed to heat and acidic conditions. In this context, note that high concentrations of formic acid, which is an excellent solubilizer [79] can cause peptide bond cleavage, particularly at aspartic acid-proline sequences [80–82], as well as formyl esterification at serine and threonine [83]. Trifluoroacetic acid, which is commonly used as a mobile phase modifier in reversed phase liquid chromatography (RPLC), is also known to accelerate aspartic acid-proline cleavage, as are other strong acids. Urea, sometimes used to solubilize proteins, can degrade to cyanate with subsequent carbamylation of the N-terminus of the protein [84]. Less well known, but still of concern, is the ability of urea to carbamylate free sulfhydryl groups [85] and other nucleophiles, including active site residues. Solvents and the surfaces that come into contact with a sample can promote other chemical modifications of amino acid side chains. The acid-catalyzed oxidation of the methionyl side chain of parvalbumin was believed to occur during chromatography [86]; oxidation of methionine is, with difficulty, reversible [87]. Other moieties, such as neuraminic acid and other glycosyl groups, may also be labile under harsh chromatographic conditions [88, 89].

Thermal Degradation

Thermal degradation is not distinct from chemical degradation. Rather, thermal elevation tends to accelerate degradative processes. One study of the thermal degradation of bovine pancreatic ribonuclease A [90] indicated that peptide bond hydrolysis at aspartic acid and deamidation at asparagine and glutamine are the principal degradation pathways at low pH. At higher pH, deamidation at asparagine and glutamine, disulfide bond reshuffling, and β-elimination of cysteine to dehydroalanine predominate. Deamidation at asparagine can proceed through a rearrangement to β-aspartate [91]. The β-aspartate rearrangement product may be stable and chromatographically separable from the unrearranged product [92].

HPLC OF PEPTIDES, POLYPEPTIDES, AND PROTEINS
General Considerations

Modern protein HPLC covers a wide variety of different column chromatographic separation methods. All of the classical column technologies have been transmuted into high performance methodologies, and most of the relevant columns are now commercially available. There are a number of reasons behind the popularity of HPLC as a separation technique. Among these are the relatively short separation times (1 to 120 min), the high sensitivity, as well the variety and availability of different detectors (e.g., absorbance, fluorescence, radioactivity, and refractive index) for detecting amounts of material in the microgram range or below. The categorization and nomenclature of HPLC methodologies is evolving perhaps more rapidly than the science, and numerous synonyms and confusing abbreviations appear in the literature. We try to use the most recent nomenclature, based on the apparent mechanism of chromatographic separation, as shown in Table 4.

Chromatographic techniques can be characterized as interactive or noninteractive depending on the presence or absence of an adsorption-desorption equilibrium. Subclassification of interactive HPLC techniques is based on the type of physical interaction involved. It is important to note that this categorization is oversimplified, and that the mechanism of a given solute-stationary phase interaction cannot be ex-

Table 4 Abbreviations Commonly Used in Modern HPLC

Type of solute-stationary phase interaction	Name	Abbreviation
None	Size exclusion chromatography	SEC
Hydrophobic	Reversed phase liquid chromatography	RPLC
	Hydrophobic interaction chromatography	HIC
Electrostatic	Electrostatic interaction chromatography, e.g.,	EIC
	ion exchange chromatography (anion and cation) and chromatofocusing	IEC
Specific	Metal chelate interaction chromatography	MIC
Specific	Biospecific interaction chromatography, affinity chromatography	BIC

pected to be rigorously homogeneous. In general, many physical forces are involved in a retention mechanism, but in practice one of these forces plays a dominant role in a given chromatography. The type and strength of the interaction of a stationary phase with a protein or peptide is a strong function of the parameters of the mobile phase; we shall return to this point later. In a categorization of chromatographies by the mechanism of analyte-stationary phase interaction, hydrophobic interaction chromatography (HIC) and RPLC are close relatives, because the mechanism of separation in both techniques is based on hydrophobic interactions.

This section focuses on the general problems of protein chromatography, with special emphasis on the use of reversed phase liquid chromatography. Reversed phase liquid chromatography is perhaps the most popular HPLC technique. Because of its apparent simplicity, versatility, and efficiency in protein and peptide isolation and analysis, RPLC has rapidly established a leading role in the biotechnology industry. RPLC of proteins, however, often involves phenomena not observed in RPLC of conventional pharmaceuticals. These include unpredictable variations of retention with alterations in protein sequence, unusual elution effects caused by solvent induced conformational alterations of denaturation, alterations in elution due to metal ions [93], column conditioning phenomena (including peak distortion, the appearance of ghost peaks, and carryover), as well as losses of activity. To some extent, such phenomena may be observed in any kind of chromatography, but these phenomena are especially common in RPLC because it relies on very strong solute-stationary phase interaction. The result is that mass transfer effects often become dominant due to the narrow range of solvent composition in which desorption occurs.

Undesired Protein Adsorption

System Involvement in Adsorption: Carryover and Recovery

Modern HPLC systems are constructed from stainless steel tubing, column, and frits. The origin of the use of metal parts is historical. In the "iron age" of chromatography, it was necessary to use parts made of steel to operate at high system pressures. The "biocompatibility" of the chromatographic system became an issue in the recent past when HPLC was applied to research problems involving sensitive and labile biomolecules. Two major problems were ascribed to HPLC parts made of stainless steel: ion leaching and protein adsorption. Leaching of metal ions from the components of the system made of stainless steel and the consequent enrichment of the mobile phase and stationary phase in metal ions may alter the chromatographic profiles and retentions of proteins. The other major problem was that of the nonspecific adsorption of protein to stainless steel parts. This can now be avoided by using tubes and columns made of noncorrosive, low adsorption titanium alloys, frequently used as biocompatible materials in medicine, or glass-lined tubing, which has recently become available. Frits, where the major undesirable adsorption occurs, can be replaced with screens of reduced surface area to minimize protein adsorption.

Column Conditioning

It has been observed in different laboratories that the height of a chromatographic peak of the components may vary in successive injections, and large differences in peak height can be observed in the consecutive chromatograms. Some proteins, generally the more hydrophobic ones, do not appear in the earlier chromatograms, while a dramatic change in the peak height can be observed for others. Eventually, with additional injections, the peak height reaches a steady state. It has been also observed that the retention time does not change as a function of the number of injections. This phenomena, known as column conditioning, is due to differences in the specific adsorption characteristics of the individual proteins. Column conditioning is found to be a general phenomenon; if, however, a new protein is then injected, additional column conditioning may be required, and some proteins idiosyncratically require more conditioning than others.

The change in peak height over successive injections can be attributed to the presence of active regions on the stationary phase: the existence of residual silanol groups as a consequence of bonding heterogeneity or generated by partial breakage of the silica beads due to the packing procedure. These "hot spots" must be saturated with protein before consistent and meaningful chromatograms can be obtained. The coverage of all available surfaces, including the tubes and frits, must also be accomplished. Different proteins saturate the chromatographic system at different rates; insulin > lysozyme > ribonuclease > albumin. It is possible that conditioning is due to the filling of the different pores, more likely the small ones, and may to a certain extent be due to irreversible denaturation and adsorption that covers the stationary phase with denatured protein. In this case adsorbed protein could then serve as a new site with a different affinity for fresh proteins. Small peptides, which are far less likely to exhibit slow denaturation, are chromatographically less problematic than proteins. Since column conditioning is a phenomenon not fully understood, and most likely results from multiple causes, columns should be conditioned when they are new, each time a new protein is analyzed, and if the analytical sample of interest is of limited quantity.

Stationary Phases; Behavior of Silica and Bonded Phases

The basis for separation of peptides and proteins is the differential adsorption of the sample components to the stationary phase. The following characteristics of the stationary phase, not necessarily in order of importance, have a special role in the separation of macromolecules:

Support (typically silica gel) type
Particle size and particle size distribution
Pore size and pore size distribution
Identity of the bonded moiety
Bonding chemistry
Coverage
Column packing procedure

HPLC of Proteins and Peptides

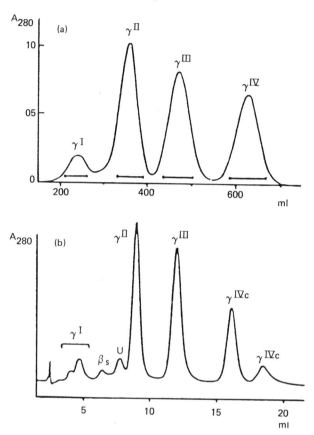

Figure 1 Comparison of high performance and conventional cation IEC of γ-crystallins. Both separations were performed in acetate buffer at pH 5.0. The vertical axes are absorbance at 280 nm, and the horizontal axes are elution volume in mL. (a) SP-Sephadex, 0.5 mL/min, 70 mg sample applied. Separation time was 7 hr. (b) Synchropak CM 300, 1 mL/min, 40 µg sample applied. Separation time was 20 min. (From Ref. 94, used with permission).

The physical strength, stability, relatively simple bonding chemistry, and long experimental experience with silica make it the most commonly used support for bonded stationary phases. The performance of silica-based stationary phases is far better than that of carbohydrate gels. Better resolution, even at higher flow rates, and significantly shorter separation times have been achieved. A comparison of separations by low pressure and high performance cation exchange methods is presented in Figure 1 [94]. Virtually identical separations were achieved in 20 min on HPLC, versus 7 hr by low pressure methods.

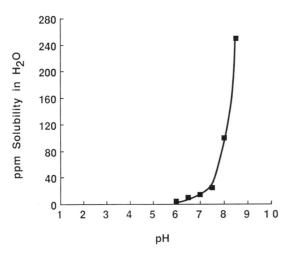

Figure 2 Solubility of amorphous silica as a function of pH. (From Ref. 95, used with permission.)

An apparent disadvantage of the silica-based stationary phases is the solubility of silica at high pH [95]. Silica-based stationary phases cannot be cleaned under the highly basic conditions biochemists have used to wash dextran-based columns. Instead, acidic conditions, which would dissolve the dextran-based stationary phases, are appropriate for cleaning silica-based stationary phases. The solubility curve of silica is shown in Figure 2 [95]. However, modifications in the technology of silica manufacturing, and the development of better bonding chemistry, resulted in commercially available bonded stationary phases with good pH and temperature stability. Some of the physical parameters of silica, such as particle size, pore size, pore size distribution, and surface area, are critical in determining the quality of the silica.

Recently, numerous new polymeric stationary phases have become commercially available. In such areas as ion chromatography and amino acid separations, polymer columns are already frequently used. Promising new phases have also been developed for peptide chromatography. An interesting application, shown in Figure 3 [96], uses a polymer-based ion exchange stationary phase for the separation of hemoglobin isoproteins. One advantage of the polymeric supports is their tolerance of a wide range of pH. Some polymers, however, shrink and swell as a function of organic solvent content, which excludes them from consideration in gradient RPLC. A recently commercialized macroporous polystyrene-divinyl benzene nonbonded phase exhibits the solvent resistance and rapid mobile-stationary phase transfer kinetics required for stable, high speed RPLC separations, even from complex mixtures such as cell culture media.

Historically, it was desirable to use small pore size silica with bonded alkyl, aryl, and cyano groups for the separation of small peptides. Large pore size stationary

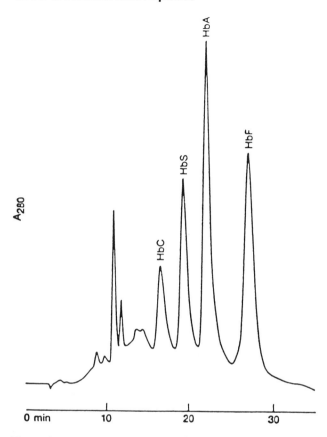

Figure 3 Separation of hemoglobin variants (HbA, HbC, HbF, and Hbs) by ion exchange on a Hydrophase HP-PEI column. Chromatographic conditions: A: 10 mM Tris, pH 8.06; B: 0.5 M NaCl. The gradient was from 0 to 40% B over 30 min. The flow rate was 1.0 mL/min. Detection was at 280 nm. The injection volume was 100 µL. (From Ref. 96, used with permission.)

phases are required for protein separations, wherein large molecules (see Table 5 [95]) must be separated from one another. The 100 Å diameter pore previously used limited analyte penetration [97–100]. The stationary phase presently most commonly used is a butyl chain bonded to a 300 Å pore size silica. Silica of pore size 300–1500 Å allows maximum penetration and interaction of proteins up to a molecular weight of approximately 150 kD. Many studies have been performed on the role of the stationary phase in the reversed phase separation of proteins. The influences of n-alkyl chain length [101–103] and silica type [104] on protein resolution, loadability, and recovery have been explored. Apparently the retention time of proteins, under gradient elution conditions, do not depend on the length of the bonded

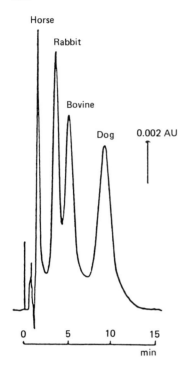

Figure 4 Isocratic separation of species variants of cytochrome c on a C_{18} reversed phase column. Chromatographic conditions: The mobile phase consisted of a mixture of 69% of 0.1 M sodium sulfate in phosphate buffer (5 mM, pH 3.0) and 31% acetonitrile. The flow rate was 2.0 mL/min. Detection was at 210 nm. The total sample size was 6 µg distributed among the species variants as 3:3:6:8 in order of elution. (From Ref. 106, used with permission.)

Table 5 Morphology, Molecular Weights, and Principal Radii of Biopolymers

Name	Morphology	Molecular weight (D)	Principal radius (ô)
Serum albumin	Solid sphere	66,000	29.8
Catalase	Solid sphere	225,000	39.8
Myosin	Rodlike	493,000	468
DNA	Rodlike	4,000,000	1170
Bushy stunt virus	Solid sphere	10,600,000	120
Tobacco mosaic virus	Rigid rod	39,000,000	924

Source: From Ref. 95, used with permission.

HPLC of Proteins and Peptides

alkyl chain. The same retention behavior has been observed in the case of C_4, C_8, and C_{18} alkyl phases, using n-propanol or acetonitrile containing eluents. However, it is important to note that significant differences have been observed in terms of the protein recovery [105].

Elution, Mobile Phases, and Recovery

Isocratic Elution

Some peptide and protein mixtures can be eluted from a reversed phase column by isocratic elution, i.e., under conditions of constant mobile phase composition and elution strength. In high performance SEC and affinity chromatography, elution typically occurs under isocratic conditions, while in RPLC, HIC, EIC, and chromatofocusing, gradient elution is more typical. Figure 4 shows that mixtures of very similar proteins can be eluted isocratically by RPLC [106]. Typical elution curves of molecules with different sizes, using isocratic reversed phase elution conditions, are shown in Figure 5 [106]. It can be seen that as the size of the molecule increases, the slope of the k' versus mobile phase concentration curve increases, and consequently the range of concentration of mobile phase organic modifier over which elution occurs decreases.

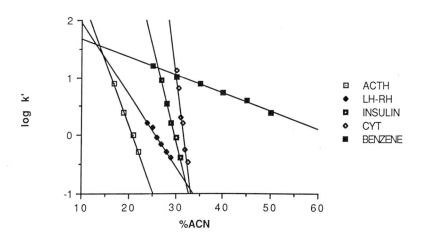

Figure 5 Dependence of the capacity factor (k') of different compounds on the acetonitrile content of the mobile phase. Chromatography was performed on a 100 × 4.6 mm id Nucleosil 7C_{18} column, using a flow rate of 1.0 mL/min. The aqueous component of the mobile phase was 0.1 M sodium sulfate in 5 mM phosphate (pH 3.0). (From Ref. 106, used with permission.)

Gradient Elution

Complex peptide and protein mixtures containing components with considerably different k' values (at any given elution concentration) cannot be separated within a reasonable time and efficiency by isocratic elution. To resolve both early and late eluting components equally well in the same chromatographic run, and to elute them as sharp, symmetrical peaks, the eluent composition must be changed continuously. The programmed change of mobile phase composition as a function of time is known as gradient elution. Routine use of gradient elution has an important practical advantage, namely that it can prevent the enrichment of the stationary phase with solvent impurities and denatured proteins. This regular cleanup increases the lifetime of the column and the reproducibility of the analysis.

Gradient elution is used in almost all types of chromatography, including ion exchange, reversed phase, and hydrophobic interaction chromatography. Good solvent mixing and reproducible gradient formation are mandatory in gradient elution. Generally, a linear gradient is most commonly used for elution, but most commercially available gradient programmers are capable of generating a variety of gradient profiles.

Mobile Phases

The roles of mobile phase parameters such as the organic modifier, pH, and ion pair reagents on peptide resolution are now well documented [107]. The most popular modifiers of the organic phase in reversed phase chromatography of proteins and peptides have been methanol, acetonitrile, n-propanol, and i-propanol. Studies have been undertaken to examine the variations in mobile phase composition on retention and recovery [83,108,109]. The addition of salts to n-propanol-and acetonitrile-based mobile phases have been shown to affect the recovery of some model proteins, and some typical results are listed in Table 6 [105]. A great influence on the protein recovery has been observed depending on the type of organic modifier applied. Propanol usually permits higher protein recovery than acetonitrile.

Generally, protein recovery drops dramatically as retention increases. More hydrophobic proteins also tend to exhibit lower recoveries. The use of mixed organic solvents may have distinct advantages over mobile phases containing only a single organic component, as is shown in Table 7 [83]. Mixtures of acetonitrile with propanol and propanol with 2-methylbutanol have also been proposed [83,110]. Mobile phase conditions such as pH, aqueous buffer type, and additives all have significant effects on protein separation as well as peak shape. In Figure 6 [105], the dramatic effects of salt concentration and type of salt additive are displayed.

Many mobile phase parameters govern the separation of proteins. The most important parameters, which a chromatographer should have knowledge of, are the following:

pH
Ionic strength
Dipole moment

Table 6 Relative Protein Recovery as a Function of *n*-Alkyl Chain Length and Organic Modifier

Column	Mobile Phase B	Relative recovery (%)				
		RNase	Cyt	BSA	CHTG	OVA
C_4	I	100	100	100	100	100
	II	79	74	69	46	35
	III	100	92	88	57	44
C_8	I	100	100	100	100	100
	II	61	68	52	54	29
	III	100	100	88	66	32
C_{18}	I	93	96	85	97	86
	II	82	75	75	57	39
	III	84	93	61	53	30

Mobile Phase A: 10 mM H_3PO_4, pH 2.2. Mobile Phase B: I = 80% *n*-propanol, 10 mM H_3PO_4; II = 80% acetonitrile; III = 80% acetonitrile, 10 mM H_3PO_4.
Source: From Ref. 105, used with permission.

Surface tension
Temperature

In the development stage of a separation, these parameters and their effect on a specific stationary phase should be evaluated and optimized. We will return to the role of mobile phases on the elution and recovery of proteins shortly.

Table 7 Recovery of Bovine Outer Segment Glycoproteins from Reversed Phase HPLC[a]

	Relative recovery (%)	
Solvent	Total glycoproteins[b]	Rhodopsin[c]
Acetonitrile	22	0
1-propanol	41	13
Acetonitrile:1-propanol	100	100

[a]A Vydac C_{18} column (4.6 × 50 mm) was used in these experiments.
[b]Based on the sum of the areas of all components recovered.
[c]Based on the area of the rhodopsin component.
Source: From Ref. 83, used with permission.

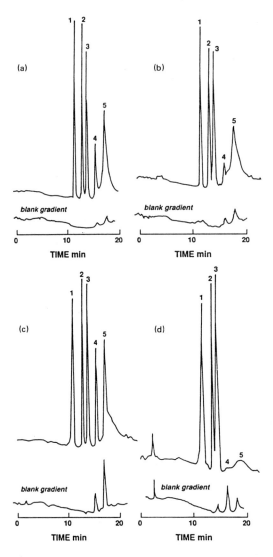

Figure 6 The effect of salt type and concentration on the reversed phase separation of proteins. The column was 100 × 4.6 mm id packed with C4 LiChrospher SI 500. Separations were performed at pH 2.2 and 37°C, with a flow of 1 mL/min. Under each chromatogram is shown a blank gradient obtained subsequent to the separation, illustrating carryover. The proteins are 1, ribonuclease A; 2, cytochrome c; 3, bovine serum albumin; 4, chymotrypsinogen; and 5, ovalbumin. The upper panel compares the effect of increasing the ionic strength of perchlorate from 16 mM (a) to 26 mM (b). The lower panel compares the effect of increasing the ionic strength of phosphate from 45 mM (c) to 205 mM (d). (From Ref. 105, used with permission.)

Separation Modes

Size Exclusion Chromatography (SEC)

Size exclusion chromatography was the first column chromatographic technique used in a high performance mode. In size exclusion chromatography (SEC), no adsorption occurs. The separation is based strictly on a simple sieving mechanism. Column materials for SEC have been improved significantly since the introduction of crosslinked dextran gels. The most popular size exclusion materials are crosslinked dextran, polyacrylamide and agarose, polymethacrylate, polyvinyl alcohol, bare and bonded silica gel, and porous glass. The difficulties in early work were largely related to unwanted protein adsorption and the physical and chemical properties of silica-based materials, as well as imperfections in the bonding chemistry. The technique has been substantially improved by the development of new bonded phase chemistry and better understanding of the retention mechanism. Lately, several new high performance SEC column materials have been developed, and excellent resolution and recovery can be now achieved with high performance size exclusion chromatography. Calibration curves are usually linear over a certain molecular weight range, indicating an ideal size-and shape-dependent distribution mechanism [111,112].

Some interaction of solute and stationary phase can occur in size exclusion chromatography if the proper mobile phase is not selected. In Figure 7, lysozyme elution is shown as a function of mobile phase salt concentration [113]. Lysozyme is not retained on the column between 0.5 and 1.5 M sodium chloride concentration, which is thus a good ionic strength range for SEC. Outside this range, lysozyme is retained on the column, with electrostatic and/or hydrophobic forces apparently dominating the adsorption mechanism. It is clear from Figure 7 that a size exclusion column can also be used for HIC. Similarly, some HIC columns can be used for SEC under appropriate conditions of ionic strength and composition of the mobile phase.

Electrostatic Interaction Chromatography (EIC)

Electrostatic interaction chromatography is frequently used for protein and peptide separations. EIC is based on electrostatic interaction between the complementary surface of the protein and the charged moieties of the stationary phase. There are two distinct types of EIC, namely IEC and chromatofocusing. The interactions between the charged stationary phase and the hydrophilic and charged surface amino acids of proteins are typically reversible and nondenaturing. Anion exchange chromatography [114,115] is more often used than cation exchange at neutral pH because the majority of proteins have isoelectric points below pH 7 [116-118].

Electrostatic interaction chromatography is ideal for detecting deamidation, differences in sialylation and phosphorylation, and posttranslational processing of the primary sequence. High performance ion exchange chromatography is also attractive because it has a resolving power which can be, in certain cases, comparable to that of electrophoresis; high recoveries are the norm (>90%). A large variety of high performance ion exchange moities are commercially available as bonded, silica-based stationary phases.

The importance of ion exchange chromatography in resolving closely related polypeptides is illustrated by the example of Figure 8 [119]. The RPLC separation of diphtheria toxin A failed to resolve three polypeptides differing only at their carboxy termini. The use of a strong anion exchanger (Mono Q) for ion exchange chromatography gave rapid, baseline resolution. Ion exchange chromatography has also been used for the purification and fractionation of the bovine pituitary peptides [120]. Ion exchange has an advantage for the chromatography of microheterogeneous proteins because certain kinds of heterogeneity are resolved by cation exchangers but not by anion exchangers, and vice versa. The rate of deamidation of a protein heterogeneous in its basic residues, for example, could be studied on an anion exchange column. The resolving power of ion exchange chromatography has been compared with that of isoelectric focusing [121].

In chromatofocusing, a complex mixture of buffers (Ampholytes) is used to generate a continuous pH gradient. Proteins bound to an ion exchange matrix elute as the gradient nears their respective isoelectric points. Chromatofocusing is an alternative to ion exchange chromatography [119] as shown in Figure 8. Chromatofocusing has some advantages over isolectric focusing, because the separation is rapid, UV detection is linear (although detection is problematic below 280 nm because of the absorbance of Ampholytes), and because recovery of high molecular weight proteins can be more readily accomplished. Chromatofocusing is an excellent sepa-

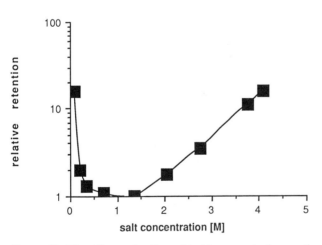

Figure 7 The effects of sodium chloride concentration on the retention behavior of lysozyme on a size exclusion column with an acetamide stationary phase. On the horizontal axis is plotted the molarity of sodium chloride in the mobile phase, while on the vertical axis is plotted the ratio of the exclusion volume (V_e) to the lysozyme retention volume (V_m). The mobile phase contained 10 mM acetate, pH 5. (From Ref. 113, used with permission.)

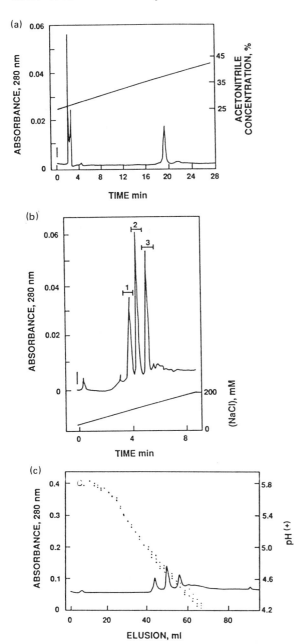

Figure 8 Chromatography of diphtheria toxin (DTA) fragments. (a) RPLC on a 4.6 × 250 mm C4 column; (b) IEC on a Mono Q column; and (c) Chromatofocusing on a Mono P column. (From Ref. 119, used with permission.)

ration method with good reproducibility, which provides the investigator with the value of the isoelectric points of the separated peptides.

Hydrophobic Interaction Chromatography (HIC)

This technique has only recently been adapted to high performance methodology, but it has the potential for numerous industrial applications. Hydrophobic interaction chromatography is carried out on weakly hydrophobic columns in high salt concentrations of ammonium or sodium sulfate [122-127]. The high surface tension of a protein dissolved in these solutions provides the solvophobic force for protein adsorption [128]. Proteins are then eluted with a descending salt gradient, and the individual proteins elute according to their solubility. One of the advantages of HIC is that biological activity is generally recovered quantitatively. Retention of proteins in HIC is extremely sensitive to the hydrophobicity of the stationary phase, which can be easily modified by both the length and the density of the alkyl side chains of the bonded phase. The capability of synthesizing stationary phases with different hydrophobicities opens up new possibilities in the optimization of separations. Changing the alkyl or aryl group attached to the hydrophilic base group affords control over retention and different selectivities, as seen in Figure 9[129]. Column regeneration is far more rapid than in RPLC. The high salt concentration increases the enrichment of trace impurities on the stationary phase, and as a consequence it increases the possibility of protein denaturation after extended use. Samples eluted from HIC columns contain high concentrations of salt, which has to be removed in a consecutive separation step (RPLC, SEC). Other chromatographic parameters, such as pH, type of salt, and temperature, can be utilized in the optimization of separations [130].

In general, based on the limited data available in the literature, hydrophobic interaction chromatography permits good, essentially quantitative sample recovery.

Reversed Phase Chromatography

RPLC, which is the most popular HPLC separation mode, is based on the hydrophobic interaction between the solute and the stationary phase. An important advantage of RPLC is that the commonly used solvent systems are volatile and can therefore be removed prior to amino acid analysis, Edman sequencing, or mass spectrometry.

A variety of alkyl and aryl chains have been thoroughly studied as bonded phases. Experiments showed that the retention as a function of n-alkyl chain length appeared constant for both propanol and acetonitrile [98,101,105]. However, as we will show later, the butyl (C_4) bonded phase seems to have become the favorite stationary phase for protein separations, because better recovery can be achieved with it. Comparisons of columns from various vendors, as shown in Figure 10 [131], indicates that column-to-column differences cause variations in chromatographic profile, resolution, and recovery [131]. The selection of the stationary phase is usually based on the user's experience with a certain type of column and/or literature precedents. The effects of the type of the silica material and its pore size on the separation of proteins have been investigated [105].

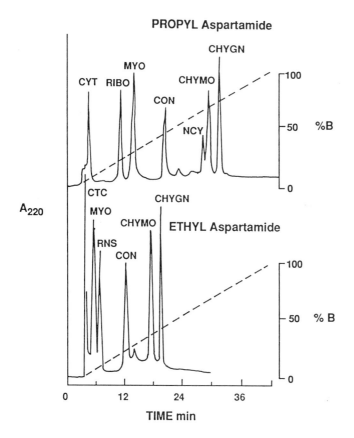

Figure 9 The effect of alkyl chain length on the separation of proteins in HIC. The retention behavior of cytochrome c, RNase A, myoglobin, conalbumin, neochymotrypsin, α-chymotrypsin, and α-chymotrypsinogen A on poly (alkyl aspartamide)-bonded silica. The gradient was from 1.5 M to 0 M ammonium sulfate, using a buffer of 100 mM phosphate, pH 7.0. The flow rate was 1 mL/min, and detection was at 220 nm. The upper panel shows elution on the propyl-aspartamide-bonded phase, and the lower panel shows elution on the ethyl-aspartamide-bonded phase. (From Ref. 129, used with permisssion.)

The application of RPLC in the area of protein and peptide separation and characterization has proven to be extremely useful. Since RPLC retention characteristics can be very sensitive to minor variations in sequence, the protein chemist interested in primary structure determination pays close attention to peak resolution. The substitution of a single amino acid residue can generate chromatographically resolvable species [106,132]. Other potentially resolvable species are glycosylation variants and species differing in the state of oxidation of the amino acid side chains [133]. The presence of disulfides can also be detected and has been employed to determine the disulfide bridging in proteins including interferon [134], T4 lysozyme [135], and tryptic fragments of human growth hormone [136].

The strong multisite binding of proteins to the n-alkyl surface and the relatively harsh conditions for protein elution, i.e., low ionic strength, low pH, and high organic solvent concentration, are known to facilitate denaturation [137]. However, for certain proteins, the recovery of the biological activity is possible [93,138,139]. In some cases, fast renaturation of the protein can occur after the chromatographic separation [140].

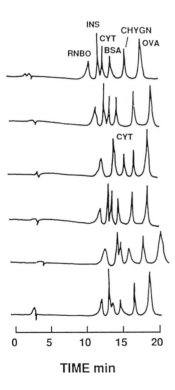

Figure 10 Comparison of retention behavior on C_8 RPLC columns from various vendors. The proteins used for the comparison were RNase A (RNS), insulin (INS), cytochrome c (CYT), bovine serum albumin (BSA), chymotrypsinogen (CHYGN), and ovalbumin (OVA). (From Ref. 131, used with permission.)

Protein Conformation and Chromatography

The adsorption of proteins at the solid-liquid interface is the fundamental phenomenon governing interactive chromatography [140-142]. As a result of adsorption, a region of the protein might interact with the stationary phase to the exclusion of other regions. Taking a rigid model of a protein molecule, it is easy to visualize that steric factors might limit the binding of the stationary phase to a surface on the protein that is complementary to the stationary phase. For example, an exposed hydrophobic region of a protein might preferentially, relative to the hydrophobic core, bind to a hydrophobic stationary phase.

The strength and type of physical forces involved in adsorption will depend on the amino acid composition of the complementary surface of the protein. The distribution of the side chains located on the exterior of the three-dimensional structure is manifested in the great variability and diversity of adsorption. The polymeric nature of proteins is manifest in phenomena such as the cooperativity of folding and unfolding. As a result of cooperativity, the total free energy of binding can be very large, even though individual site-site interactions may be weak.

Studies attempting to establish a relationship between retention time and hydrophobic amino acid content of polypeptides or proteins have shown very little correlation of these parameters. Efforts to make retention time predictions in reversed phase elution have failed for polypeptides with more than 20 residues [143,144], demonstrating that the size and three-dimensional structure of proteins play a governing role in the adsorption process, and emphasizing the importance of a complementary surface.

The crucial role of surface complementarity has been shown in the HIC, EIC, and metal affinity chromatography of lysozyme [145,146]. Some characteristics of the complementary surface can be measured by using the stoichiometric displacement model (SDM) [130,147,148]. The SDM assumes that a solvent molecule or ion adsorbed to a definable class (D_b) of sites is displaced by chromatographically competitive functional groups of surface side chains of the protein P_0 according to a stoichiometry Z to generate free solvent molecules or ions D_0 and bound protein P_b according to the equation

$$P_0 + ZD_b = P_b + ZD_0$$

The capacity factor k' is related to the stoichiometric coefficient Z of solvent molecules or ions that are displaced, and to the concentration [D_0] of the solvent, and is expressed by

$$k' = \frac{I}{[D_0]^Z}$$

or

$$\log k' = \log I - Z \log[D_0]$$

where I is a constant (with a defined and important, but complicated, physical meaning). The Z number can be determined by graphing log k' versus the concentration of the relevant mobile phase ion or solvent molecule, and the slope of the curve is then -Z and the intercept is log I. The SDM approach has proven useful for the characterization of protein adsorption in ion exchange [149], reversed phase [147], and hydrophobic interaction [130] chromatography. Although the displacement process is slightly different in each chromatographic mode, the same equation can be applied to all of them.

The analysis of Z numbers derived under a variety of different conditions can illuminate the adsorption mechanism of a protein. Typical Z values of proteins under hydrophobic interaction chromatographic conditions, as shown in Figure 11 [130], are indicative of a multiple binding point mechanism of adsorption. Because desorption of proteins occurs over a very narrow range of the eluent concentration, a large Z value also indicates a high level of cooperativity between these individual attachment points.

The effects of temperature on the retention of proteins have been shown to correlate with the Z number in hydrophobic interaction chromatography of α-lactalbumin, γ-lactoglobulin, and cytochrome c [130]. These studies showed that Z increased with temperature. Using spectroscopic methods, the Z number increase was demonstrated to be related to the geometry of the complementary surface of the protein. For the analysis of genetically engineered biomacromolecules, where the identification of minor modifications in the protein structure cannot usually be detected by conventional spectrophotometry, the existence of a method to differentially detect folding-unfolding transitions is valuable. Cytochrome c has a highly conserved primary structure; Table 8 [150] shows the differences between the amino acid sequences of cytochromes extracted from various biological sources. As is shown in Figure 12 [106], RPLC can separate species variants of cytochrome c, using isocratic elution on a C_8 column. Five species variants of cytochrome c were successfully separated, while bovine and chicken coeluted. Comparison with Figure 4, which shows isocratic RPLC of cytochrome c species variants on a C_{18} column, demonstrates the importance of the stationary phase in determining the separation. Reversal of the elution order was observed on a cyanoalkyl phase. The ability to obtain a separation was ascribed to differences in the conformation caused by the amino acid substitutions, rather than to the side chain hydrophobicity.

Retention Change on Denaturation

Variations in protein chromatographic retention can be caused by disruption of the native conformational equilibrium. The main effects that can cause these changes are listed in Table 9. Under denaturing chromatographic conditions, one must contend with the occurrence of almost all of these effects. Some of the conformational

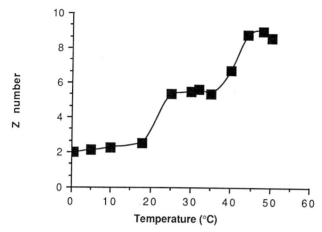

Figure 11 The Z value of α-lactalbumin on HIC as a function of temperature. On the horizontal axis is the elution temperature in °C, and on the vertical axis is the calculated Z value. Sharp increases in the slope on increasing temperature are indicative of thermal unfolding. (From Ref. 130, used with permission.)

Table 8 The Amino Acid Sequences of Species Variants of Cytochrome c[a]

Position	Bovine	Horse	Dog	Rabbit	Chicken	Tuna
4	Glu	Glu	Glu	Glu	Glu	*Ala*
9	Ile	Ile	Ile	Ile	Ile	*Thr*
15	Ala	Ala	Ala	Ala	*Ser*	Ala
22	Lys	Lys	Lys	Lys	Lys	*Asn*
28	Thr	Thr	Thr	Thr	Thr	*Val*
33	His	His	His	His	His	*Trp*
44	Pro	Pro	Pro	*Val*	*Glu*	*Glu*
46	Phe	Phe	Phe	Phe	Phe	*Tyr*
47	Ser	*Thr*	Ser	Ser	Ser	Ser
54	Asn	Asn	Asn	Asn	Asn	*Ser*
58	Thr	Thr	Thr	Thr	Thr	*Val*
60	Gly	*Lys*	Gly	Gly	Gly	*Asn*
61	Glu	Glu	Glu	Glu	Glu	*Asn*
62	Glu	Glu	Glu	*Asn*	*Asp*	*Asp*
88	Lys	Lys	*Thr*	Lys	Lys	Lys
89	Gly	*Thr*	Gly	*Asp*	*Ser*	Gly
92	Glu	Glu	*Ala*	*Ala*	*Val*	*Gln*
95	Ile	Ile	Ile	Ile	Ile	*Val*
100	Lys	Lys	Lys	Lys	*Asp*	*Ser*
103	Asn	Asn	Asn	Asn	*Ser*	*Ser*
104	Glu	Glu	Glu	Glu	*Lys*	—

[a]Residues that differ are indicated in italics.
Source: From Ref. 150, used with permission.

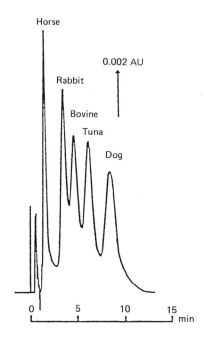

Figure 12 Separation of species variants of cytochrome c on RPLC. The stationary phase was C_8, and the mobile phase was 72.5% of 0.1 M sodium sulfate in 5 mM phosphate (pH 3.0) and 27.5% acetonitrile. The flow rate was 2.0 mL/min. Detection was by 220 nm absorbance. The total sample size was 11 µg, with a distribution of 2:2:2:2:3 of the species indicated in the order of elution. Comparison with Figure 4 demonstrates the importance of the stationary phase in determining the characteristics of the separation. (From Ref. 106, used with permission.)

Table 9 Environmental Parameters Affecting Protein Conformation

Parameter	Typical Effect
Temperature	Denaturation at elevated temperature
Solvent	Lowering of transition temperature inorganic and chaotropic solvents
pH	Denaturation on protonation of carboxyl groups (acidic pH) or deprotonation of tyrosyl protons (basic pH)
Interfaces	Denaturation, especially at liquid-solid and gas-liquid interfaces due to competition of intra-and intermolecular forces

HPLC of Proteins and Peptides

changes observed upon adsorption are gentle and reversible. The protein retains conformational properties close to those in the solution state, and elutes as biologically active material. Other conformational changes on adsorption may be extensive, leading to significant alteration in the thermodynamic and biological properties of the molecule. Direct observation of quasireversible pH denaturation of sperm whale apomyoglobin adsorbed to a C_{18} phase has been reported [151]. Protein denaturation may be even more facile under chromatographic conditions as compared with conditions of dilute aqueous solution, since denaturation due to adsorption and denaturation due to solvent effects can be synergistic [137].

It has been observed that under reversed phase and hydrophobic interaction chromatographic conditions, the denatured conformer elutes later in the gradient than the native protein [130,140,141,152,153]. However, in ion exchange and affinity chromatography, the opposite elution behavior was observed [154]. This phenomenon can be explained by the change of the complementary surface. In the case of RPLC and HIC, the effective area of the complementary surface increases on denaturation, providing more attachment points and stronger binding. In EIC, denaturation dilutes the local concentration of charged groups, reducing the cooperativity of binding.

The Effects of Disulfide Reduction

The reduction of the disulfide bonds greatly destabilizes the native structure by cleavage of the only covalent chemical bond responsible for the stabilization of long-range interactions in the native protein. As one might expect, the chromatographic behavior of a disulfide-containing protein is altered. Figure 13 [79] displays the chromatograms of a numerous proteins before and after disulfide reduction with tri-n-butylphosphine (TBP). Cytochrome c, whale myoglobin, horse myoglobin, carbonic anhydrase, and ovalbumin seem to exhibit little shift in retention on reduction. RNase A, lysozyme, BSA, conalbumin, trypsin inhibitor, and β-lactoglobulins A and B exhibit greater retention after disulfide reduction. Reduction of the disulfides therefore increases the effective hydrophobic contact area between the protein and the stationary phase.

The Effect of Organic Modifiers

Organic solvents used in RPLC also affect the conformation of proteins and accordingly their elution. These effects can be manifested in differences in retention time, in peak shape, or in the number of peaks. In Figure 14 [141], the effects of acetonitrile, 2-propanol, and 1-propanol on trapping a nondenatured form of α-chymotrypsin have been compared. The injections were made immediately or 9 min after the gradient was started. It is clear that α-chymotrypsinogen denatured extremely rapidly when the mobile phase was acetonitrile, so the only form observed was the denatured form. Denaturation was slower in 2-propanol, and much slower in 1-propanol.

Temperature Induced Conformational Changes

The effect of column temperature on the chromatographic behavior of proteins has been investigated. Using soybean trypsin inhibitor [141], papain [141,155], lysozyme [130,156,157], cytochrome c [130,156], myoglobin [156], and RNase A

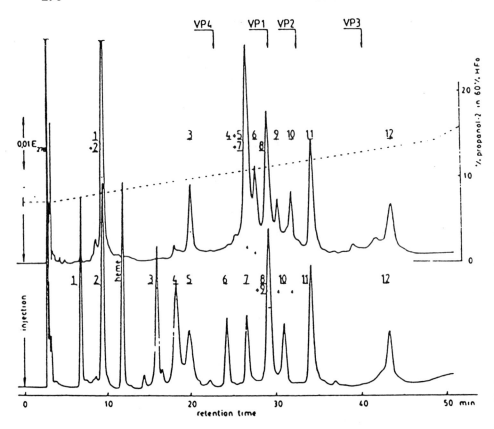

Figure 13 Comparison of the chromatographic behavior on gradient RPLC of disulfide-reduced (upper panel) and disulfide-intact (lower panel) proteins. Reduction was accomplished with TBP. The separations were performed on an Aquapore RP-300 column in 60% formic acid, using a gradient with 2-propanol. The proteins used were 1, RNase A; 2, horse cytochrome c; 3, hen egg lysozyme;l 4, BSA; 5, bovine conalbumin; 6, trypsin inhibitor; 7, whale myoglobin; 8, horse myoglobin; 9 and 10, β-lactoglobulin A and B; 11, bovine carbonic anhydrase; and 12, ovalbumin.

[140,158] as the model proteins, it was demonstrated that folding-unfolding transitions can occur dynamically during chromatography. Native and denatured conformational states can create two or more peaks or distorted peaks. The height of the first chromatographic peak, which corresponds to the native state, decreases as the temperature increases.

HPLC of Proteins and Peptides

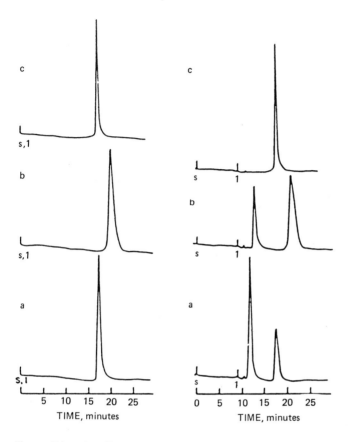

Figure 14 The effect of the organic modifier and the injection time on the gradient reversed phase chromatography of α-chymotrypsinogen. The stationary phase was LiChrospher Si 500, an endcapped C4 phase. The aqueous mobile phase was 10 mM H_3PO_4, pH 2.2. The organic mobile phase was (a) 1-propanol:water in a 45:55 v/v ratio; (b) 2-propanol:water in a 45:55 v/v ratio; or (c) acetonitrile:water in an 80:20 v/v ratio. The overall H_3PO_4 concentration was 10 mM. The chromatograms on the left are of α-chymotrypsinogen injected immediately after the start of the gradient, while the chromatograms on the right were obtained by injecting α-chymotrypsinogen 9 min after the start of the gradient. The peak eluting later in the gradient is the denatured form.

Kinetics Effects of Denaturation on Chromatographic Elution

The kinetics of denaturation during chromatography have been studied in certain model systems [155,157]. Using soybean trypsin inhibitor [141,159], papain [141,155], lysozyme [141,155], and RNase A [140,158], it was demonstrated that both irreversible and reversible denaturation can occur during chromatography. Native and denatured conformational states often elute with different retention times [152,155,159], reflecting the changes in the complementary surface of the protein, and this has been used to determine the kinetics of denaturation under chromatographic conditions [141,157]. The ability to distinguish between the native and the denatured protein conformation is a function of the kinetics of the conformational equilibrium. The presence of peaks corresponding to the native and denatured conformational states depends on the competition between the kinetics of the conformational change and the velocity of elution [139,140,160,161]. Distorted peaks were observed when the half-life of refolding was comparable to the time scale of chromatography.

Cofactor and Prosthetic Group Induced Conformational Alterations

Calcium-binding proteins such as calmodulins and parvalbumins elute as broad peaks under normal elution conditions [93]. These proteins elute as sharp, well-defined peaks when $CaCl_2$ or ethyleneglycol-bis-(β-aminoethyl ether)-N,N,N',N'-tetraacetic acid (EGTA) is added to the mobile phase. As the pH of the eluent is lowered from 7 to 2.5, calcium is released from its binding site on the protein, and the retention time increases. It is likely that the structure of the apoprotein differs from that of the calcium-bound protein, due to pH induced denaturation upon the release of calcium. The increase in retention time reflects the additional attachment points between the analyte and the stationary phase available to the apoprotein, resulting both from denaturation and from the protonation of carboxyl groups, which has been shown to increase retention in reversed phase separations [144]. Similar changes in the retention behavior have been observed in the HIC separation of α-lactalbumin as a function of mobile phase calcium concentration. In order to obtain sharp peaks, the appropriate cofactor or chelating agent must be added to the mobile phases.

An example of the effect of side chain modification on RPLC retention of disulfide-reduced bovine serum albumin is shown in Figure 15 [79]. Alkylation of the cysteinyl residues with different reagents, which alters the complementary surface and/or overall conformation of the protein, has a remarkable effect on the retention characteristics of protein.

Microheterogeneity

The substitution of one amino acid for another at a single position of the primary sequence of a protein is known to occur during protein synthesis [106,132]. Other

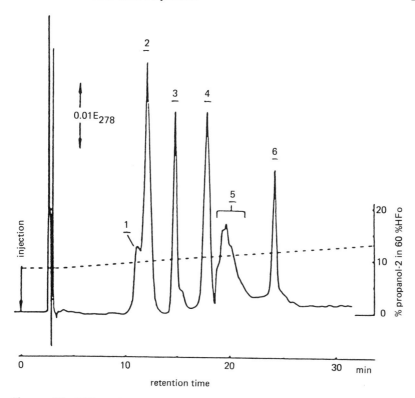

Figure 15 RPLC of disulfide-reduced, cysteine-blocked BSA. The column was an Aquapore RP 300. Gradient elution was performed in 60% formic acid, with a positive gradient of 2-propanol. The BSA was reduced with dithioerythritol and cysteine blocked with the following groups: 1, aminoethyl; 2, pyridyl; 4, carboxyamidomethyl; and 5, sulfoethyl. Unreduced BSA, 3, and Disulfide-reduced BSA, 6, are also shown. (From Ref. 79, used with permission.)

detectable chemical modifications include glycosylation, which may be the most prevalent form of microheterogeneity, and oxidation of the labile amino acid side chains [133]. Glycosylation variants in t-PA have been detected by electrophoretic methods [71,162] and further characterized by RPLC peptide mapping/Edman sequencing [163]. Bovine pancreatic ribonucleases A (nonglycosylated) and B (glycosylated at Asn 35) can be separated by RPLC and by cation exchange chromatography [119]. The extent of deglycosylation on treatment of RNase B with α-mannosidase, endoglycosidase H, or endoglycosidase D can be followed by high performance IEC, as shown in Figure 16 [119]. The chromatographic resolution of the glycosylated and deglycosylated forms of RNase is much greater than the electrophoretic resolution of the same preparation.

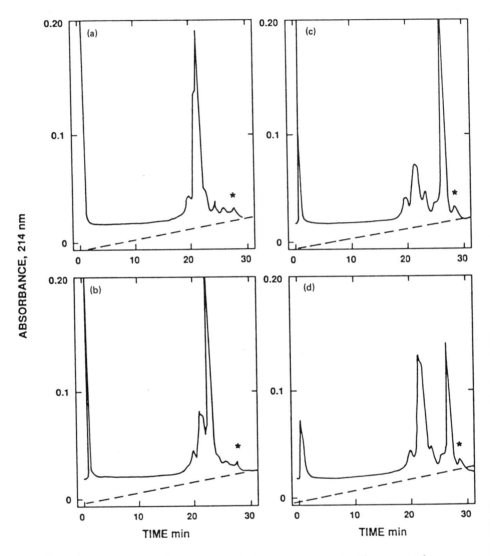

Figure 16 Cation IEC of untreated and deglycosylated RNase B. Chromatography was performed on a Mono S column, using a buffer of 25 mM phosphate (pH 6.0) and a linear gradient of 0–1 M sodium chloride. The flow rate was 1 mL/min, and detection was by absorbance at 214 nm. (a) No treatment; (b) α-mannosidase treatment; (c) endoglycosidase H treatment; and (d) endoglycosidase D treatment.

The presence of disulfides can also be detected chromatographically. This principle has been employed to determine the occurrence of disulfide bridging in proteins such as interferon [134], T4 lysozyme [135], and by peptide mapping of disulfide-intact human growth hormone [136]. Other forms of protein microheterogeneity can include phosphorylation of serine and threonine, deamidation, and oxidation of methionine.

PROTEIN HPLC IN THE PHARMACEUTICAL INDUSTRY

General

High performance liquid chromatography has been central in the field of pharmaceutical science, and pivotal in the biopharmaceutical arena. Specific and sensitive analytical methods have become extremely important in confirming product identity, quality, and lot-to-lot reproducibility in production. Polypeptides with almost identical chemical compositions can exhibit similar or utterly different biological activities; so separation scientists involved in various aspects of the isolation and characterization of a polypeptide product of pharmaceutical interest may have to be able to separate those entities. On the other hand, production chemists may wish to suppress the separation of pharmacologically equivalent variants to maximize the yield of biologically active compound. The development and production of any compound for pharmaceutical uses requires that separations be performed from complex, often poorly characterized matrices, and production and analysis are complicated by chemical and enzymatic degradation that occur during handling.

Issues

HPLC is used in almost every area of drug discovery and development as well as in the production process. The development of HPLC methods in the production of peptide and protein pharmaceuticals centers around the following fundamental topics:

Production and the automation of production
Validation of the production process
Determination of identity and concentration
Preformulation analysis
Stability testing
Analysis of purity
Specifications testing
Quality control
Physicochemical characterization

HPLC in Production

In the pharmaceutical industry, the primary goal of the production staff is to obtain the maximum yield of the desired molecule(s) in a pharmacologically active form with a minimal expenditure of resources. The existence of chemically similar variants of biopharmaceutical drug substances often makes purification and analytical testing peculiarly challenging. This is especially the case in the early stages of development, when the natures of "inactive" species are unknown. Sample stabilities, solubilities, and routes of degradation are also largely unknown in the early developmental stages. Chromatographic methods are used in production to establish the identity and concentration of the desired product in a sample.

In planning the production and characterization of a protein product, it is important to know the pathways and products of degradation and the nature of the host proteins that are likely to co-migrate with the protein of interest on a given separation system. A knowledge of the pathways and products of degradation, if obtained in the early stages of development, can be useful in planning the purification scheme and in guiding stability studies [164]. Because deamidation is an important pathway of degradation, high performance IEC (anion exchange) or another charge-sensitive separation method might be considered as a production diagnostic. It is also wise to determine which impurities are proteolytically active to ensure that these impurities are removed or destroyed early in the production process [165]. If possible, the final purification step should not involve exposure of the product to a pH less than 4 or greater than 8, because outside of this range chemical degradation is accelerated. It has been pointed out [166] that considerations of scale make the typical purification scheme one of precipitation, followed successively by ion exchange, affinity, and gel filtration chromatography. A review of the recovery of recombinant proteins from *E. coli* [167] discusses production issues in biopharmaceutical compounds generated by this producer type.

There has been increased regulatory interest in using process validation, in addition to the required postproduction analytical testing, as a means of building quality control into production. Process validation quantitates the extent of removal of an impurity of interest by a particular process step. In process validation, it is especially important to analyze the sensitivity of the process to variations in the scale of the separation, to modifications of production methods, and to process repetition. Just as column age, sample loading, and variations in the sample matrix can cause changes in resolution and retention in analytical chromatography, these factors can cause degradation of process performance. Another approach to quality control in production is the examination of lot-to-lot variations in trace impurities (impurity profiling). Impurity profiling is potentially a powerful diagnostic tool.

Analytical Principles in Biopharmaceuticals

Selection of the Reference Standard and Design of the Testing Program

One of the production batches is reserved as the reference standard. Its composition is determined using all appropriate methodologies, and subsequent production batches are assayed against it. Assays should be tailored to answer definable issues, such as the removal of a known process chemical, or, for a chemically modified product, the specificity and lot-to-lot uniformity of the modification. Because of problems of matrix interference following formulation, some analyses are best per-

HPLC of Proteins and Peptides

formed on the bulk drug, i.e., before formulation. Most proteins decompose at measurable rates even under controlled, refrigerated storage, so considerable judgment is required in comparative assays for the determination of purity. During the characterization of the reference standard, those assays that are believed to be essential to the characterization of subsequent batches are examined in detail and thereby validated.

Test Validation

The analytical parameters that are most important in test validation and subsequent analysis are detection limits, range of assay linearity, run-to-run variability (precision), day-to-day variability (reproducibility), consistency of the results of one test with those of other tests, sample stability, ruggedness (a definition of conditions under which a test may fail), and system specificity (a definition of those impurities that a given separation system is capable of resolving). During the subsequent analysis of the first few production batches, which are typically intended for preclinical testing in animal models, the production lots are assayed against the reference standard to establish specifications for the drug substance. Once specifications are established, it is crucial that the production process be well enough controlled that subsequent production batches meet those specifications, so that clinical trials, which must be scheduled well in advance, can proceed without delay. For similar reasons, ruggedness is a central consideration in the choice of analytical methods, often requiring that one choose a relatively insensitive but reliable method over one that is more sensitive, but prone to variability.

Specifications

Close collaboration between those responsible for the purification of a compound and those responsible for its analysis is necessary to establish specifications that can be met under the conditions of production. Specifications, which define minimal criteria by which a drug substance may be released for preclinical and clinical use, must be based on simple, reliable tests that are diagnostic of definable issues relating to safety and efficacy. This point cannot be overemphasized. Each protein is unique, so the analytical tests should target specific issues, such as aggregation, deamidation, and lot-to-lot variation. One cannot perform all possible tests, so the choice of which tests to use should be a carefully reasoned choice.

The Formulated Drug Substance

As part of the process of drug formulation, certain assays critical to the guarantee of the stability and efficacy of the formulated drug are developed and similarly validated. The development of such assays is typically complicated by the sample matrix of formulation additives. Therefore, the presence of trace process chemicals and other trace contaminants is examined prior to formulation, and issues of stability dominate the analytical process after formulation. Formulated samples are subjected to a series of stressed conditions (elevated temperatures, exposure to light, and exposure to extremes of pH) to determine the proper storage conditions and the acceptable shelf life of the biopharmaceutical. The knowledge of the pathways of degradation so determined is potentially of great utility not only in the formulation and storage of the product, but also in the optimization of the purification scheme: one of the characteristics of an effective development staff is the rapid feedback of stability data to the process chemists.

The first few batches of the drug substance are used in the development of an acceptable formulation, and tested in animal models. At this stage, only the simplest assays of composition and activity are required, but much more sophisticated assays to determine the pathways of biological clearance must be developed. At the same time, the framework must be laid to establish the chemical basis of the correspondence of the animal model to the human response that will be observed in subsequent clinical trials.

Identity, Homogeneity, and Purity

Issues in the Definition of Identity

The first step in the analysis of a sample is to establish that it is essentially homogeneous according to explicit criteria. The criteria by which it is deemed to be homogeneous (such as the observation of a single major chromatographic peak at a given retention time) can constitute indirect proofs of identity. Unfortunately, even the definition of homogeneity can become a perplexing issue, given the kinds of heterogeneity that can be exhibited by recombinant products. For protein and peptide pharmaceuticals, the initial assays for homogeneity are typically performed by SDS-PAGE and RPLC. Because the purification process used for production is typically optimized using these methods, subsequent analysis should, when practical, include orthogonal methods. A precise definition of orthogonality would require that the fraction capacity of the analytical process should increase the fraction capacity of the joint production-analytical processes by an amount approaching the product of the fraction capacities of the individual processes [168,169]. For practical purposes, orthogonal methods can be conceived to be methods that differ substantially in a mechanistic sense from those used for purification or for process optimization.

It is not always possible to use orthogonal methods in analysis, particularly if the production process involves numerous kinds of chromatography. Of course, if the production process uses numerous kinds of chromatography, the risk that a significant amount of any given impurity will persist in the product is decreased. If the production process is an abbreviated one, that risk is increased. The use of identical methods for steps of purification and for quality control of the drug substance is, therefore, a dubious practice, likely to produce inflated estimates of purity, and carrying the potential of compromising the safety and efficacy of the product.

Proof of Structure

As has been noted, errors in DNA transcription can cause the translated protein sequence to differ from that of the DNA. Such errors may be subtle, making the determination of the identity of a drug substance difficult. The determination of identity may be difficult for other reasons. For example, it is becoming common practice in molecular biology to ligate an invariant portion of DNA to a portion that one varies to produce constructs with very similar properties (the "cassette approach"). As sensible as this approach may be from the standpoint of drug design, it increases the difficulty in devising appropriate tests of identity. Only by completing a rigorous characterization of the reference standard one can establish the validity of indirect proofs of identity (such as migration on SDS-PAGE, antibody reactivity on Western Blotting, or chromatographic retention) for the characterization of subsequent samples.

One can define a general strategy for the characterization of the reference standard. The *N*-terminus is a site of active processing, and therefore of potential heterogeneity. Accordingly, the *N*-terminus must be maximally sequenced, not only to establish the identity of the protein but also to determine the extent of *N*-terminal heterogeneity. The carboxy terminus is also sequenced, typically by analysis of the amino acids released by carboxypeptidase treatment, but the difficulties of sequencing the carboxy terminus are far greater than those of sequencing the amino terminus. The identity of the reference standard must be further established by obtaining a full sequence of the sample by peptide mapping of the disulfide-reduced protein and Edman sequencing or fast atom bombardment (FAB) mass spectrometry [170,171] of the fragments. FAB mass spectrometry is particularly useful in identifying peptide fragments with covalently attached prosthetic groups; when the *N*-terminus is blocked to Edman sequencing, mass spectrometric techniques are essential for the confirmation of the sequence [172]. FAB methodology has been very helpful in the analysis of the oligosaccharide location, composition, and structure in glycoproteins. In tPA, for example, treatment of a tryptic digest of reduced, carboxymethylated tPA with peptide:*N*-glycosidase F, an endoglycosidase that hydrolyzes the β-aspartylglycosylamine linkage, resulted in the appearance of signals corresponding to residues 113–129, 177–189, and 441–449 [173,174]. The appearance of signals not observed prior to peptide:*N*-glycosidase F treatment indicated that residues within these sequences are glycosylated in the native protein. The disulfide bond pattern of the reference standard should be determined by peptide mapping of the disulfide-intact protein. If the drug substance has been deliberately chemically modified, peptide mapping of the disulfide-reduced form can help to identify the site and extent of modification.

Characterization of the Reference Standard

In addition to the proof of structure, the properties of the reference standard should be thoroughly determined. The protein concentration should be established by amino acid analysis or, if the buffer does not interfere, by micro Kjeldahl analysis [175]. From the concentration as determined by amino acid analysis, an indirect method for the determination of concentration, such as spectrophotometric absorbance or RPLC, can be calibrated. Also, the purity of the reference standard must be determined by chromatographic and electrophoretic methods, so that subsequent decomposition on storage can be detected. A preliminary catalog of trace impurities should be established for comparison with future batches. The amount of each process chemical in the sample, or an upper bound of its concentration, must be established. Occasionally, a process chemical will interfere with the performance of another assay. If an assay for a process chemical relies on spiking samples of the drug substance with known amounts of that chemical, the concentration of the analyte in the reference standard should be well below the specification limit. Process chemicals include the components of cell culture, DNA, and materials that leach from columns during purification.

Batch Analysis

Peak retention on RPLC, amino acid analysis, Western Blotting, fingerprint peptide mapping [176], and migration on SDS-PAGE under reducing and nonreducing conditions are relatively rapid secondary proofs of identity and purity that can be repeated on each production batch. The reference standard is used for the validation of these methods, and to check day-to-day assay performance. Typically, each production batch is sequenced for 20 or more residues. If the operation of the sequencer has been validated with the reference standard, batch-to-batch variations in proteolytic processing can be quantitated. The reference standard is also useful in comparative batch analysis of residual process chemicals.

Practical Issues in Chromatographic Batch Analysis

For an analytical method to be useful in specifications testing, the detection limits for individual impurities must be well below the total impurity limit defined by the specifications. If, for example, specifications require a purity limit of 99.0% by a given method, the detection limits for that method should be well below the 1% level. By the time a drug substance enters analytical testing, the levels of individual impurities are likely to be at or below parts per thousand. Therefore, the emphasis in purity analysis is on the quantitation of trace components. The detection limit of an incompletely resolved impurity is a benchmark of the performance of an analytical method. If no impurities can be detected by an analytical method, the resolution and detection limits of the method are probably suboptimal.

Even carefully validated assays can sometimes fail, so it is important to establish a daily record of assay performance. For purity determinations, it is desirable to use an instrument dedicated to specifications testing. An instrument used for specifications testing must have a log of calibrations, tests, and instrumental repairs. The system suitability should be established on a daily basis by the analysis of standards. Standards can also be used to verify the completeness of column conditioning prior to the injection of samples. Blank injections are useful. Injection of the analyte buffer, diluted with the same buffer used to dilute samples, helps to identify artifacts traceable to the sample matrix and to sample preparation. The injection of a blank after the injection of a sample may help to establish the extent of carryover between samples. The inclusion of the reference standard in a series of chromatographic runs provides a direct check on the performance of the chromatographic and the data analysis systems. In analyzing chromatographic results, overlaying expanded replicate runs with a blank can help to differentiate genuine peaks from noise; blank subtraction has the disadvantage that noise increases. The baseline used for determination of purity should coincide with a blank obtained by diluting the analyte buffer with the same proportion of sample dilution buffer used for preparing actual samples. This is because it is not uncommon that numerous trace components elute essentially continuously over a portion of the chromatogram, resulting in a very broad, flat impurity peak, difficult to recognize as such without careful comparison of expanded overlays of sample and blank chromatograms.

The data system used for specifications testing must be validated. Although computer validation is a complex issue, three points are of immediate interest to the practicing chromatographer. First, the exact location of integration start and stop and the position of the baseline used for integration should be clearly indicated on the chromatogram. Second, it should be verified that the density of data points is adequate to the analysis. Third, the precise effects of alterations in the parameters used for integration should be systematically investigated.

Sources of Protein Impurities

To decide which of the protein species should constitute the pharmaceutical product may not be as simple as it might seem. With tPA, for example, species including glycosylation variants, the 1-and 2-chain forms, and the L-and S-chain forms all might be included in the drug substance. One might wish, however, to exclude species that exhibit proteolysis at sites other than the L-, S-, and 2-chain forms. It is necessary to reach consensus on which of the many species to include and which to exclude from purity calculations before it is possible to examine the impurity profile of the protein.

Assuming that the structure of a variant protein is very similar to that of the desired product, the purification process might fail to separate it. If the alteration involves a change in an ionizable residue, techniques such as isoelectric focusing, chromatofocusing, and ion exchange chromatography might separate the variant. Alternatively, the presence of a variant protein might be detectable by enzymatic digestion followed by peptide mapping and sequencing of the fragments. Other sources of heterogeneity, however, are frequently encountered. The processing variants of tissue plasminogen activator were discussed previously; variants known as L-and S-chain and 1-and 2-chain are known. No differences of therapeutic importance are known for the four variants, and resolution is difficult, so the separation of these variants must be suppressed during production. For the characterization of clinical supplies, however, it is important to characterize the sample as fully as possible, in case some therapeutic advantage or disadvantage is later discovered. Heterogeneity involving proteolysis may be detected by separation after disulfide reduction; peptide mapping/Edman sequencing of the isolated fragments is used to localize the site of cleavage [163]. In addition to these forms of heterogeneity, the presence or absence of cofactors, the degree of aggregation of multisubunit proteins, and conformational differences can cause two preparations to behave heterogeneously in a therapeutic sense, even though techniques such as amino acid analysis, SDS-PAGE, Edman sequencing, and peptide mapping would deem them to be entirely homogeneous.

Even if the DNA sequence has been carefully verified, it should never be assumed that a protein produced by cell culture is the desired one. As has been pointed out above, errors in protein synthesis can occur due to a variety of factors. It is even possible for large segments of DNA to be mistranscribed. Development of reliable, reproducible instrumental assays indicative of conformational integrity and activity is definitely an important priority on the analytical agenda.

HPLC in the Determination of Purity

The most important parameters that must be considered in developing an HPLC purity assay are:

Recovery
Linearity
Detection limit
Peak integrity

Because of the inherent complexity of biopharmaceutical products, it is wise to examine column conditioning by using a new column. Low recoveries are expected during the first few runs and may be detected by examining the total peak area during column conditioning. This phenomenon has been discussed in detail under Undesired Protein Adsorption and under Stationary Phases: Behavior of Silica and Bonded Phases.

Linearity tends to be the least important of these parameters, partly because detection with UV detectors is well understood, and partly because an assay must be well out of linear range before the apparent purity will be seriously affected; in such a case the apparent purity of the principal component would be expected to be underestimated rather than overestimated. Linearity studies, however, are important in establishing the range of loading under which optimal quantitation is attained. Graphing the apparent purity as a function of sample load is often helpful in establishing the optimal range.

The detection limit is typically established from the linearity data obtained on the principal component. Because the response factors for impurities are typically unknown, results are reported in area percent. Often, materials eluting at the void volume are ignored, but one can examine the early part of the chromatogram by lyophilizing the sample and redissolving it in the starting mobile phase. It is improper to dismiss peaks detected at void volume as artifacts without performing the necessary control experiments.

Peak integrity is the most difficult issue in the method validation procedure. Peak integrity, which defines the resolving power of the chromatographic method, can be examined spectrophotometrically, chromatographically, and by two-dimensional methodologies. Well-developed spectroscopic methodologies permit the characterization of proteins and the detection of protein conformational changes [177–179]. Using diode array detection, a wide variety of these methods can be utilized for purity determinations, as indicated in Table 10.

The absorbance spectra at the chromatographic peak maximum and at the ascending and descending half-heights are often compared as a proof of peak integrity. Using a variety of schemes involving wavelength ratios at the absorbance maxima of the three aromatic amino acids (Figure 17), peak integrity determination can be

Table 10 Spectroscopic Methods for the Assessment of Peak Integrity

Method	Detection mode	Peak location	Wavelength (nm)
Wavelength ratioing	Multiple wavelength diode array detection	Apex	Trp: 254, 274, 292; Tyr: 254, 274; Phe: 254
Difference Spectroscopy	Diode array detection	Apex	Full spectrum
Derivative spectroscopy	Diode array detection	Apex	Full spectrum

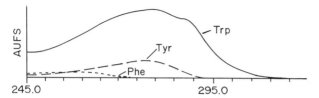

Figure 17 UV spectra of equimolar solutions of N-AcTrpNH$_2$, N-AcTryNH$_2$, and N-Ac-PheOEt in 0.5 M ammonium acetate, pH 6. Spectra were recorded on a Hewlett-Packard Model 1040A photodiode detector. (From Ref. 130, used with permission.)

made more sensitive [180]. Derivative spectroscopy has also been used for peak purity determination. Of course, conformational transitions may interfere with the determination of peak integrity [130,150,158].

Chromatographic characteristics, as described in earlier sections, can also be used to establish peak integrity. The experiments involved in preparing plots of k' versus percentage organic modifier, which are constructed from a series of isocratic elutions, help to identify the optimal conditions to resolve components which coelute on gradient elution. Because the retention times of various proteins are nonidentical functions of the temperature, changes of the retention time, peak width, or peak symmetry with changes in column temperature may be of help in assessing peak integrity. The use of similar stationary phases prepared by different vendors may result in elution profiles with slightly or even drastically different characteristics. This can be problematic in comparing data from different laboratories.

Changes of retention time, peak width, and peak symmetry as a function of column temperature can be observed in the chromatograms of α-lactalbumin as presented in Figure 18 [130]. Figure 18 plots the retention volume of α-lactalbumin as a function of the column temperature. Note that the curve is discontinuous, indicating a conformational change.

On of the most successful two-dimensional approaches is to combine RPLC with SDS-PAGE. Contaminants will often partially resolve from the major peak on RPLC, even though resolution is not detectable by inspection of the chromatogram. By loading relatively large amounts of cuts obtained from the peak edges onto SDS-PAGE, the intensities of the contaminants on silver staining can be increased manyfold. This approach, which has been used in our laboratory [164], has an additional advantage, namely, that it can be used to differentiate artifactual PAGE bands, which will be observed continuously across the peak, from genuine contaminant bands.

HPLC in Protein Characterization

Amino Acid Analysis

In amino acid analysis, a protein or peptide is hydrolyzed by strong acid or base, and many of the difficulties with amino acid analysis are caused by hydrolysis problems and not by chromatographic factors. Hydrolysis is normally accompanied by appre-

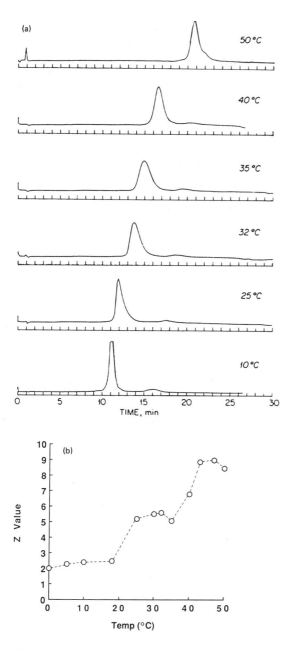

Figure 18 The effect of temperature on the chromatographic characteristics of α-lactalbumin on HIC. (a) Chromatograms of α-lactalbumin eluted from an ether-bonded phase in a buffer of 0.5 M ammonium acetate, pH 6, using a gradient from 2 to 0 M ammonium sulfate at a flow rate of 1.0 mL/min. The temperature is indicated on the chromatogram. In (b) is shown the effect of the temperature in °C (horizontal axis) on the corrected retention volume V_g. (From Ref. 130, used with permission.)

ciable degradation of glutamine, asparagine, tryptophan, cysteine, cystine, methionine, tyrosine, serine [181], and threonine. Glutamine and asparagine rapidly deamidate to glutamic acid and aspartic acid, respectively. Glutamic acid and serine can condense to O-(γ-L-glutamyl)-L-serine [182]. Methionine can be oxidized during hydrolysis; methionine sulfoxide is not stable to hydrolysis, but methionine sulfone is stable [183]. Cystine can degrade to a number of products, including cysteine [184,185] and bis-(β-amino-β-carboxyethyl)trisulfide [186]. Serine and threonine exhibit essentially linear decay [187]. Timed hydrolysis is useful in determining the completion of hydrolysis and the extent of hydrolytic losses. Although the principal cause of oxidation in amino acid analysis is atmospheric oxygen, there is evidence [188] that oxygen can be chemically incorporated into HCl, the most commonly used hydrolytic agent, perhaps as HOCl and higher oxides. Tactics in limiting degradation during amino acid analysis include alkylation of the cysteine and cystine residues and hydrolysis in the presence of antioxidants, such as thioglycollic acid [189,190]. Problems with oxidation become especially serious in analyzing small samples. Exogenous amines, notoriously Tris, are common interferents.

Separations for underivatized amino acids are conventionally performed using ion exchange chromatography [191]. The underivatized amino acids can be visualized by postcolumn derivatization with ninhydrin [192], fluorescamine [193], or o-phthaladehyde [194,195]. The o-phthalaldehyde system has been modified to include detection of proline [196].

If the amino acids are derivatized prior to chromatography, reversed phase chromatography is most often used for the separation. Reagents that have been developed for precolumn derivatization include 9-fluorenylmethyl chloroformate [197], dansyl chloride, i.e., 1-N,N-dimethylaminonaphthalene-5-sulfonyl chloride [198,199], dabsyl chloride, i.e., 4,4-dimethylamino-4'-benzenesulfonyl chloride [200], o-phthalaldehyde/2-mercaptoethanol [201], and phenylisothiocyanate [202].

N-Terminal Sequencing

Automated Edman sequencing is used for determining the N-terminal sequence. The vast amount of literature in this area has been recently reviewed [203]. In this method, the protein is immobilized with a hydrophilic polymer. The N-terminus is coupled to phenyl isothiocyanate, and the derivatized amino acid is cleaved with strong acid and converted to the phenylthiohydantoin derivative, which is analyzed by RPLC. Serine and threonine generate additional components, namely dehydroalanine, dehydrothreonine, and adducts with dithiothreitol. Cysteine is not observed. Numerous by-products also form, and much of the literature is devoted to analytical approaches to reducing by-product interferences. A practical guide to the operation of the gas phase sequencer produced by Applied Biosystems is available in the literature [204].

Peptide Mapping

In peptide mapping, the protein of interest is typically reduced and S-carboxymethylated, and the carboxymethylated form is treated with a protease or a chemical hydrolytic agent to generate peptide fragments. These are separated by HPLC and indi-

HPLC of Proteins and Peptides

vidually identified by amino acid analysis or sequencing. Two-dimensional peptide mapping [205] has not gained widespread acceptance in the pharmaceutical industry, because the data are tedious to interpret. This is unfortunate, because two-dimensional chromatographic peptide mapping is potentially a powerful diagnostic tool that is much more reliable than two-dimensional gel methods. Peptide maps are intrinsically complex, and therefore the choice of the hydrolysis method for generating the peptide fragments is crucial. Trypsin and cyanogen bromide are the preferred hydrolytic agents because of their high specificities. Careful control of all of the experimental conditions is essential to obtaining reproducible peptide maps. Because of the complexity of and variability in peptide maps, multidimensional detection is valuable. The simple use of simultaneous 280 nm and 220 nm detection allows one to assign peaks that contain (absorb at both 280 and 220 nm) or lack (absorb only at 220 nm) aromatic amino acid residues. If the slope of the baseline is much smaller than the peak height, it is possible to identify incompletely resolved peptides by such simple peak integrity measurements. Incompletely resolved fragments may need to be rechromatographed prior to amino acid analysis. Fragments containing easily oxidized residues may require special treatment for amino acid analysis. But the potentials for using automated postcolumn detection systems are extremely exciting. Almost any colorimetric assay can be adapted to a postcolumn system. Such a system was used for the automated detection of thiols and disulfides [206] in the peptide mapping of disulfide-intact ribonuclease. In principle, it should be possible to develop automated assays to identify glycopeptides in peptide maps [207]. In the area of protein separations, continuous postcolumn assays should also make it possible to automate activity assays, to test peak integrity, or to detect the presence of contaminating proteases.

HPLC in Stability Testing

Before a protein or peptide can be released for clinical use, its stability and solubility under various conditions must be established. Typical concerns in stability testing are the sensitivity of the pharmaceutical to photochemical degradation, to thermal degradation, and to hydrolysis under extended storage. Thermal acceleration of decomposition can be established by analyzing the purity of samples held at different temperatures as a function of time. Of course, near the transition temperature of the unfolding of a globular protein, one might expect to see changes in the pathway of decomposition as residues protected from the solvent below the transition temperature become exposed to solvent. The solubility of the product is also important in the preparation of the final dosage form, because proteins and peptides must often be injected. Obviously, one would like to minimize the injection volume to minimize patient discomfort, but the formation of aggregates may be favored by high concentration. The regulations for permissible pharmaceutical vehicles are extremely restrictive, requiring considerable ingenuity on the part of the formulation chemist.

In the analysis of the final formulated product (including the analysis of the solubility and the stability), chromatographic methods are attractive because of their rapidity and reproducibility. Given the complex degradation pathways that proteins and peptides may exhibit, and given the additional complexity of analyzing a pharmaceutical in its formulation matrix, the choice of analytical methods for a formulated product requires a thorough knowledge of biochemical characterization methods. Some of the mechanisms of degradation that merit investigation are deamidation, peptide bond hydrolysis, disulfide bond reshuffling, and aggregation. A charge-sensitive methods, such as IEC on anion exchange, may detect deamidation. Peptide bond hydrolysis may be detected by reducing SDS-PAGE or chromatography on the disulfide-reduced protein. Disulfide reshuffling may be detected by peptide mapping of the disulfide-intact protein. Because noncovalent aggregates may disperse on SDS-PAGE, SEC may be more useful in the determination of aggregation. Of course, if aggregation is the consequence of intermolecular disulfide bond reshuffling, nonreducing SDS-PAGE may be useful, too.

The design of stability protocols is highly regulated, and there are nation-to-nation differences in requirements. Although the final design of a protocol may be determined largely by regulatory considerations, it is helpful to consider the underlying analytical issues. Stability testing, by definition, takes time, so careful design of the stability protocol pays. For simplicity, consider the design of an accelerated thermal decomposition protocol. At higher temperatures, thermal unfolding may be accompanied by a change in the decomposition pathway. Therefore, predictions of shelf life are likely to be reliable only below the transition temperature. Sampling too frequently would waste both analytical time and sample, but one needs to sample frequently enough that the acceptable period of storage for clinical samples can be estimated promptly. Therefore, one would like to take samples at intervals over which the purity has decreased to an extent just discernible by the analytical method. Obviously, the use of sample replicates (or sampling frequently) and population statistics can partially compensate for deficiencies in analytical methods, but the returns diminish quickly with increasing sample load. Decisions that help to coordinate production schedules and the timetables of clinical trials can hinge on the results of stability tests, so the day-to-day reproducibility and the detection limits of the analytical method used to evaluate stability need to be substantially better than the maximum tolerable level of decomposition. One can use this design approach to detect egregious mismatch of an analytical method to the requirements of a stability protocol. For example, a method reproducible to 1% is probably not appropriate if a 1% decline in purity would be cause for concern. The day-to-day reproducibility of chromatographic methods is generally much better than that of most activity assays. Therefore, if a stability-indicating chromatographic method can be devised, a reliable prediction of shelf life can probably be made sooner than by activity assay.

Activity Analysis

The measurement of the activity of a peptide or protein is usually not based on chromatographic procedures, at least in the discovery phase. The mass requirements of the various HPLC methods are relatively large, compared to the requirements of radioimmunoassy (RIA), enzyme-linked immunoabsorbent assay (ELISA), capillary electrophoresis, and SDS-PAGE, so there seems to have been surprisingly little effort devoted to adapting automated activity assays to HPLC. The introduction of microbore HPLC technology has drastically reduced the amount of sample required, but not to the levels of alternative techniques. Later in the development cycle, as the throughput and precision of the required assays becomes as issue of primary importance (as in preformulation and stability studies), and sample consumption becomes a secondary issue, chromatography would seem to be the method of choice. There is no intrinsic reason that the colorimetric or radiometric detection systems on which ELISA, RIA, and Western Blot analysis are based cannot be adapted to chromatographic systems, and it is likely that effective automation is best done by this approach [115].

Optical Activity On Line

Detectors using polarimetric [208-210] or circular dichroic [209- 210] methods have been reported. Until recently, however, the sensitivity of such methods was relatively low. Recently, however, substantial improvements in polarimetric detectors have been reported [211-212], permitting sensitive detection of samples of biological interest. Two recent reports included determination of the specific optical rotation of numerous peptides [213,214] and proteins [214] by RPLC. Such methods may prove to be useful in determinations of peak integrity and conformational integrity. They are already finding enthusiastic acceptance in the analysis of conventional pharmaceuticals.

CONCLUDING REMARKS

The principal concerns regarding biopharmaceutical agents are those of safety and efficacy. Components of the cell culture that lack the desired activity are presumed to be deleterious, and are removed to the extent that is possible. If the active component is to be used with a broader patient population than is permitted with investigational new drugs, those components that cannot be removed to levels below available detection limits may need to be characterized or proven to be harmless. The issues of stability and solubility are of importance for all members of the development team. Therefore, close coordination between those departments involved in production, formulation, and analytical testing is crucial.

Ideally, the microscale purification procedures that are developed in the earliest stages of discovery are designed to mesh with the scaled-up processes that will be required in later stages of development. If so, there is the potential for using data from the microscale procedures to help troubleshoot, or perhaps even validate, the purification process. The benefits of changes in fermentation and purification methods should be balanced against the potential costs in analysis and batch failure. Well-reasoned, well-timed, and well-validated changes in production methods fit naturally into pharmaceutical development. The effort required to bring a biopharmaceutical compound through development is considerable, but it can be moderated by intelligent coordination of the human element in development. In the words of Cuatresecas [215], ". . . drug development cannot be managed in the traditional sense. The 'managers' must rather be strong leaders, accomplished and respected scientists themselves, who must exhibit broad vision, long-term perspective, trust in other professionals, and the ability to inspire others. . . The public and the ethical industry are best served by decision based on good science, adherence to high standards, and independent, expert review. . . If the industry starts with high quality science, effective analyses, and honest, responsive presentations, its regulatory problems will be few."

ACKNOWLEDGMENTS

We wish to express our gratitude to Dr. C. S. Randall, Dr. J. J. L'Italien, and Dr. S. A. Carr of Smith Kline and French Laboratories for helpful discussions and suggestions during the writing of this chapter.

ABBREVIATIONS

BPTI	Bovine pancreatic trypsin inhibitor
BSA	Bovine serum albumin
CHTG	Chymotrypsinogen
DNA	Deoxyribonucleic acid
EGTA	Ethyleneglycol-bis-(β-aminoethyl ether)-N,N,N',N'-tetraacetic acid
EIC	Electrostatic interaction chromatography
ELISA	Enzyme-linked immunosorbent assay
HIC	Hydrophobic interaction chromatography
IEC	Ion exchange chromatography
IEF	Isoelectric focusing
kD	Kilodaltons
LYSO	Lysozyme
OVA	Ovalbumin
RIA	Radioimmunoassay
RNA	Ribonucleic acid
RNase	Bovine pancreatic ribonuclease
RPLC	Reversed phase liquid chromatography
SDS-PAGE	Sodium dodecyl sulfate-polyacrylamide gel electrophoresis
SEC	Size exclusion chromatography
t-PA	Tissue plasminogen activator
UV-VIS	Ultraviolet-visible

REFERENCES

1. W. Szmuness, C. E. Stevens, E. J. Harley, E. A. Zang, W. R. Oleszko, D. C. William, R. Sadovsky, J. M. Morrison, and A. Kellner, *New Engl. J. Med. 303*: 833 (1980).

2. J. F. Young, W. T. Hockmeyer, M. Gross, W. R. Ballou, R. A. Wirtz, J. H. Trosper, R. L. Beaudoin, M. R. Hollingdale, L. H. Miller, C. L. Diggs, and M. Rosenberg, *Science, 228*: 958 (1985).
3. L. H. Miller, R. J. Howard, R. Carter, M. F. Good, V. Nussenzweig, and R. S. Nussenzweig, *Science, 234*: 1349 (1986).
4. W. R. Ballou, J. A. Sherwood, F. A. Neva, D. M. Gordon, R. A. Wirtz, G. F. Wasserman, C. L. Diggs, S. L. Hoffman, M. R. Hollingdale, W. T. Hockmeyer, I. Schneider, J. F. Young, P. Reeve, and J. D. Chulay, *The Lancet II*, June: 1277 (1987).
5. H. K. Webster, E. F. Boudreau, L. W. Pang, B. Permpanich, P. Sookto, and R. A. Wirtz, *J. Clin. Microbiol., 25*: 1002 (1987).
6. E. M. Bonnem, *Investigational New Drugs, 5*: S65 (1987).
7. C. E. Welander, *Investigational New Drugs, 5*: S47 (1987).
8. D. L. Howrie, *Clin. Pharmacy, 6*: 283 (1987).
9. S. J. Crabbe and C. C. Cloniger, *Clin. Pharmacy, 6*: 373 (1987).
10. I. J. Hollander, *CRC Rev. Biotech., 6*: 253 (1987).
11. P. Tiollais, C. Pourcel, and A. DeJean, *Nature, 317*: 489 (1985).
12. D. L. Peterson, *BioEssays, 6*: 258 (1987).
13. J. Pêtre, F. Van Wijnendaele, B. De Neys, K. Conrath, O. Van Opstal, P. Hauser, T. Rutgers, T. Cabezon, C. Capiau, N. Harford, M. De Wilde, J. Stephenne, S. Carr, M. Hemling, and J. Swadesh, *Postgrad. Medical J., 63(S2)*: 73 (1987).
14. J. R. Brown and B. S. Hartley, *Biochem. J., 101*: 214 (1966).
15. B. S. Hartley and D. L. Kauffman, *Biochem. J., 101*: 229 (1966).
16. B. S. Hartley, *Nature, 201*: 1284 (1964).
17. G. N. Ramachandran, C. Ramakrishnan, and V. Sasisekharan, *J. Mol. Biol., 7*: 95 (1963).
18. G. Veliçelebi, S. Patthi, and E. T. Kaiser, *Proc. Natl. Acad. Sci. USA, 83*: 5397 (1986).
19. K.-C. Chou, G. Nèmethy, S. Rumsey, R. W. Tuttle, and H. A. Scheraga, *J. Mol. Biol., 186*: 591 (1985).
20. R. S. Morgan, C. E. Tatsch, R. H. Gushard, J. M. McAdon, and P. K. Warme, *Int. J. Pept. Protein Res., 11*: 209 (1978).
21. G. Nèmethy and H. A. Scheraga, *BBRC, 98*: 482 (1981).
22. J. K. Swadesh, P. W. Mui, and H. A. Scheraga, *Biochemistry, 26*: 5761 (1987).
23. M. Lebl, E. E. Sugg, and V. J. Hruby, *Int. J. Pept. Protein Res., 29*: 40 (1987).
24. M. Lebl in *CRC Handbook of HPLC for the Separation of Amino Acids, Peptides, and Proteins, Vol. II* (W. S. Hancock, ed.), CRC Press, Boca Raton, Florida, p. 169 (1984).
25. R. Kornfeld and S. Kornfeld, *Ann. Rev. Biochem., 54*: 631 (1985).
26. I. B. Holland, N. Mackman, and J.-M. Nicaud, *Biotechnology, 4*: 427 (1986).
27. D. Barnes, J. van der Bosch, H. Masui, K. Miyazaki, and G. Sato, *Meth. Enzymol., 79*: 368 (1981).
28. J.-M. Nicaud, N. Mackman, and I. B. Holland, *J. Biotech., 3*: 255 (1986).
29. T. Imanaka, in *Advances in Biochemical Engineering/Biotechnology, Vol. 33* (A. Fiechter, ed), Springer-Verlag, Berlin, p. 1 (1986).
30. A. Bollen, *J. Biotech., 2*: 317 (1985).
31. M. G. Tovey, *Meth. Enzymol., 79*: 391 (1981).
32. D. W. Jayme and K. E. Blackman, *Advances in Biotechnological Processes, 5*: 1 (1985).
33. R. K. Scopes, *Protein Purification*, Springer-Verlag, New York, p. 25 (1982).
34. P. A. Maher and S. J. Singer, *Proc. Natl. Acad. Sci. USA, 83*: 9001 (1986).
35. V. G. Janolino, H. E. Swaisgood, and H. R. Horton, *J. Applied Biochem., 7*: 33 (1985).

36. A. Light, *BioTechniques, 3*: 298 (1985).
37. P. N. Lewis, F. A. Momany, and H. A. Scheraga, *Proc. Natl. Acad. Sci. USA, 68*: 2293 (1971).
38. D. B. Wetlaufer, *Proc. Natl. Acad. Sci. USA, 70*: 697 (1973).
39. J. F. Lefczynski and G. D. Rose, *Science, 234*: 849 (1986).
40. C. B. Anfinsen, E. Haber, M. Sela, and F. H. White, Jr., *Proc. Natl. Acad. Sci. USA, 47*: 1309 (1961).
41. P. Y. Chou and G. D. Fasman, *Advances in Enzymology, 47* (A. Meister, ed.), John Wiley, New York, p. 45 (1978).
42. W. L. Mattice and H. A. Scherage, *Biopolymers, 23*: 1701 (1984).
43. S. Tanaka and H. A. Scheraga, *Macomolecules, 10*: 291 (1977).
44. S. Tanaka and H. A. Scheraga, *Proc. Natl. Acad. Sci. USA, 72*: 3802 (1975).
45. J. E. Brown and W. A. Klee, *Biochemistry, 10*: 470 (1971).
46. D. N. Silverman, D. Kotelchuck, G. T. Taylor, and H. A. Scheraga, *Arch. Biochem. Biophys., 150*: 757 (1972).
47. A. Bierzynski and R. L. Baldwin, *J. Mol. Biol., 162*: 173 (1982).
48. M. Rico, J. L. Nieto, J. Santoro, F. J. Bermejo, J. Herranz, and E. Gallego, *FEBS Lett., 162*: 314 (1983).
49. J. K. Swadesh, G. T. Montelione, T. W. Thannhauser, and H. A. Scheraga, *Proc. Natl. Acad. Sci. USA, 81*: 4606 (1984).
50. E. R. Stimson, G. T. Montelione, Y. C. Meinwald, R. K. E. Rudolph, and H. A. Scheraga, *Biochemistry, 21*: 5252 (1982).
51. L. Pauling, R. B. Corey, and H. R. Branson, *Proc. Natl. Acad. Sci. USA, 37*: 205 (1951).
52. J. A. Reynolds, D. B. Gilbert, and C. Tanford, *Proc. Natl. Acad. Sci. USA, 71*: 2925 (1974).
53. M. F. Perutz, J. C. Kendrew, and H. C. Watson, *J. Mol. Biol., 13*: 669 (1965).
54. W. Kauzmann, *Advances in Protein Chem., 14*: 1 (1959).
55. J. B. Matthew, *Ann. Rev. Biophys. Biophys. Chem., 14*: 387 (1985).
56. M. Sundaralingam, Y. C. Sekharudu, N. Yathindra, and V. Ravichandran, *Proteins: Structure, Function, and Genetics, 2*: 64 (1987).
57. P. J. Milburn, Y. Konishi, Y. C. Meinwald, and H. A. Scheraga, *J. Am. Chem. Soc., 109*: 4486 (1987).
58. T. E. Creighton, *J. Mol. Biol., 144*: 521 (1980).
59. T. E. Creighton, *J. Mol. Biol., 129*: 411 (1979).
60. T. E. Creighton, *Meth. Enzymol., 131*: 83 (1986).
61. J. F. Brandts, H. R. Halvorson, and M. Brennan, *Biochemistry, 14*: 4953 (1975).
62. J. F. Brandts and L.-N. Lin, *Meth. Enzymol., 131*: 107 (1986).
63. Z. Li and H. A. Scheraga, *Proc. Natl. Acad. Sci. USA, 84*: 6611 (1987).
64. G. L. Reddy and R. Hagaraj, *Int. J. Pept. Protein Res., 29*: 497 (1987).
65. C. Zwizinski and W. Wickner, *J. Biol. Chem., 255*: 7973 (1980).
66. P. B. Wolfe, W. Wickner, and J. M. Goodman, *J. Biol. Chem., 258*: 12073 (1983).
67. K. H. S. Swamy and A. L. Goldberg, *Nature, 292*: 652 (1981).
68. C. C. Dykstra and S. R. Kushner, *J. Bacteriol., 163*: 1055 (1985).
69. K. H. S. Swamy and A. L. Goldberg, *J. Bacteriol., 149*: 1027 (1982).
70. J. S. Bond and P. E. Butler, *Ann. Rev. Biochem., 56*: 333 (1987).
71. P. Wallèn, G. Pohl, N. Bergsdorf, M. Rånby, T. Ny, and H. Jörnvall, *Eur. J. Biochem., 132*: 681 (1983).
72. H. Jörnvall, G. Pohl, N. Bergsdorf, and P. Wallèn, *FEBS Lett., 156*: 47 (1983).

73. P. Wallèn, M. Rånby, and P. Kok, *Prog. Chem. Fibrinolysis Thrombolysis,* 5: 16 (1981).
74. N. Sharon and H. Lis, *Nature,* 323: 203 (1986).
75. D. M. Carrington, A. Auffret, and D. E. Hanke, *Nature,* 313: 64 (1985).
76. P. H. O'Farrell, *Cell,* 14: 545 (1978).
77. J. Parker, T. C. Johnston, P. T. Borgia, G. Holtz, E. Remaut, and W. Fiers, *J. Biol. Chem.,* 258: 10007 (1983).
78. J. Parker, J. W. Pollard, J. D. Friesen, and C. P. Stanners, *Proc. Natl. Acad. Sci. USA,* 75: 1091 (1978).
79. J. Heuskeshoven and R. Dernick, *J. Chromatogr.,* 252: 241 (1982).
80. M. Landon, *Meth. Enzymol.,* 47: 145 (1977).
81. P. Sonderegger, R. Jaussi, H. Gehring, K. Brunschweiler, and P. Christen, *Anal. Biochem.,* 122: 298 (1982).
82. J. Rittenhouse and F. Marcus, *Anal. Biochem.,* 138: 442 (1984).
83. G. E. Tarr and J. W. Crabb, *Anal. Biochem.,* 131: 99 (1983).
84. G. R. Stark, *Meth. Enzymol.,* 25: 579 (1972).
85. G. R. Stark, *J. Biol. Chem.,* 239: 1411 (1964).
86. M. W. Berchtold, K. J. Wilson, and C. W. Heizmann, *Biochem.,* 21: 6552 (1982).
87. R. A. Houghten and C. H. Li, *Meth. Enzymol.,* 91: 549 (1983).
88. G. Blix and R. W. Jeanloz, in *The Amino Sugars, Vol. IA* (R. W. Jeanloz, ed.), Academic Press, New York, p. 231 (1969).
89. H. Rinderknecht and T. Rebane, *Experientia,* 19: 342 (1963).
90. S. E. Zale and A. M. Klibanov, *Biochemistry,* 25: 5432 (1986).
91. P. Bornstein and G. Balian, *Meth. Enzymol.,* 47: 132 (1977).
92. T. W. Thannhauser and H. A. Scheraga, *Biochemistry,* 24: 7681 (1985).
93. M. W. Berchtold, C. W. Heizmann, and K. J. Wilson, *Anal. Biochem.,* 129: 120 (1983).
94. R. J. Siezen, E. D. Kaplan, and R. D. Anello, *Biochem. Biophys. Res. Comm.,* 127: 153 (1985).
95. W. W. Yau, J. J. Kirkland, and D. D. Bly, *Modern Size-Exclusion Liquid Chromatography,* John Wiley, New York (1979).
96. J. R. Benson and N. Kitagawa, *J. Chromatog.* 443:133 (1988).
97. R. V. Lewis, A. Fallon, S. Stein, K. D. Gibson, and S. Udenfriend, *Anal. Biochem.,* 104: 153 (1980).
98. E. C. Nice, M. W. Capp, N. Cooke, and M. J. O'Hare, *J. Chromatogr.,* 218: 569 (1981).
99. K. J. Wilson, E. van Wieringen, S. Klauser, M. W. Berchtold, and G. J. Hughes, *J. Chromatogr.,* 237: 407 (1982).
100. J. D. Pearson, W. C. Mahoney, M. A. Hermodson, and F. E. Regnier, *J. Chromatogr.,* 207: 325 (1981).
101. R. V. Lewis and D. DeWald, *J. Liq. Chromatogr.,* 5: 1367 (1982).
102. J. D. Pearson and F. E. Regnier, *J. Liq. Chromatogr.,* 6: 497 (1983).
103. N. H. C. Cooke, B. G. Archer, M. J. O'Hare, E. C. Nice, and M. Capp, *J. Chromatogr.,* 255: 115 (1983).
104. J. D. Pearson, N. T. Lin, and F. E. Regnier, *Anal. Biochem.,* 124: 217 (1982).
105. K. A. Cohen, K. Schellenberg, K. Benedek, B. L. Karger, B. Grego, and M. T. W. Hearn, *Anal. Biochem.,* 140: 223 (1984).
106. S. Terabe, H. Nishi, and T. Ando, *J. Chromatogr.,* 212: 295 (1981).

107. M. T. W. Hearn in *Advances in Chromatography, Vol. 20* (J. C. Giddings, E. Grushka, J. Cazes, and P. R. Brown, eds.), Marcel Dekker, New York, p. 1 (1982).
108. M. J. O'Hare, M. W. Capp, E. C. Nice, N. H. C. Cooke, and B. G. Archer, *Anal. Biochem., 126*: 17 (1982).
109. W. C. Mahoney and M. A. Hermodson, *J. Biol. Chem., 255*: 11199 (1980).
110. J. P. Chang, W. R. Melander, and Cs. Horvath, *J. Chromatogr., 318*: 11 (1985).
111. H. Engelhardt, and D. Mathes, *J. Chromatogr., 142*: 311 (1977).
112. P. Andrews, *Br. Med. Bull., 22*: 109 (1966).
113. H. Engelhardt and M. Czok, Personal Communication
114. A. J. Alpert and F. E. Regnier, *J. Chromatogr., 185*: 375 (1979).
115. F. E. Regnier and K. M. Gooding, *Anal. Biochem., 103*: 1 (1980).
116. P. G. Righetti and T. Caravaggio, *J. Chromatogr., 127*: 1 (1976).
117. P. G Righetti, G. Tudor, and K. Ek, *J. Chromatogr., 220*: 115 (1981).
118. A. J. Alpert, *J. Chromatogr., 266*: 23 (1983).
119. K. J. Wilson, *User Bulletin*, Applied Biosystems; Fremont, CA p. 1 (1985).
120. S. James and H. P. J. Bennett, *J. Chromatogr., 326*: 329 (1985).
121. Y. Kato, K. Nakamura, Y. Yamazaki, and T. Hashimoto, *J. Chromatogr., 318*: 358 (1985).
122. Y. Kato, T. Kitamura, and T. Hashimoto, *J. Chromatogr., 292*: 418 (1984).
123. J. L. Fausnaugh, E. Pfannkoch, S. Gupta, and F. E. Regnier, *Anal. Biochem., 137*: 464 (1984).
124. D. L. Gooding, M. N. Schmuck, and K. M. Gooding, *J. Chromatogr., 296*: 107 (1984).
125. J.-P. Chang, Z. El Rassi, and Cs. Horváth, *J. Chromatogr., 319*: 396 (1984).
126. N. T. Miller, B. Feibush, and B. L. Karger, *J. Chromatogr., 316*: 519 (1985).
127. S. C. Goheen and S. C. Engelhorn, *J. Chromatogr., 317*: 55 (1984).
128. W. R. Melander, D. Corradini, and Cs. Horváth, *J. Chromatogr., 317*: 67 (1984).
129. A. J. Alpert, *J. Chromatogr., 359*: 85 (1986).
130. S.-L. Wu, K. Benedek, and B. L. Karger, *J. Chromatogr., 359*: 3 (1986).
131. K. A. Cohen, S. A. Grillo, and J. W. Dolan, *LC, 3*: 37 (1985).
132. P. E. Petrides, R. T. Jones, and P. Böhlen, *Anal. Biochem., 105*: 383 (1980).
133. J. E. Zull and J. Chuang, *Anal. Biochem., 140*: 214 (1984).
134. R. Wetzel, *Nature, 289*: 606 (1981).
135. L. J. Perry and R. Wetzel, *Science, 226*: 555 (1984).
136. W. J. Kohr, R. Keck, and R. N. Harkins, *Anal. Biochem., 122*: 348 (1982).
137. C. Tanford, *Adv. Protein Chem., 23*: 121 (1968).
138. K. Titani, T. Sasagawa, K. Resing, and K. A. Walsh, *Anal. Biochem., 123*: 408 (1982).
139. M. P. Strickler, M. J. Guski, and B. P. Doctor, *J. Liq. Chromatogr., 4*: 1765 (1981).
140. S. A. Cohen, K. Benedek, Y. Tapuhi, J. C. Ford, and B. L. Karger, *Anal. Biochem., 144*: 275 (1985).
141. K. Benedek, S. Dong, and B. L. Karger, *J. Chromatogr., 317*: 227 (1984).
142. F. E. Regnier, *Science, 238*: 319 (1987).
143. J. L. Meek and Z. L. Rossetti, *J. Chromatogr., 211*: 15 (1981).
144. D. Guo, C. T. Mant, A. K. Taneja, J. M. R. Parker, and R. S. Hodges, *J. Chromatogr., 359*: 499 (1986).
145. J. L. Fausnaugh and F. E. Regnier, *J. Chromatogr., 359*: 131 (1986).
146. R. Drager and F. E. Regnier, *J. Chromatog, 402*:237 (1987).
147. X. Geng and F. E. Regnier, *J. Chromatogr., 296*: 15 (1984).
148. R. R. Drager and F. E. Regnier, *J. Chromatogr., 359*: 147 (1986)
149. M. A. Rounds and F. E. Regnier, *J. Chromatogr., 283*: 37 (1984).

150. M. O. Dayhoff, L. H. Hunt, P. J. McLaughlin, and W. C. Barker, *Atlas of Protein Sequence and Structure,* Vol. 5, National Biomedical Research Foundation, Washington, D.C., p. D7 (1972).
151. C. H. Lochmüller and S. S. Saavedra, *J. Am. Chem. Soc., 109:* 1244 (1987).
152. S. Y. M. Lau, A. K. Taneja, and R. S. Hodges, *J. Chromatogr., 317:* 129 (1984).
153. A. J. Sadler, R. Micanovic, G. E. Katzenstein, R. V. Lewis, and C. R. Middaugh, *J. Chromatogr., 317:* 93 (1984).
154. E. S. Parente and D. B. Wetlaufer, *J. Chromatogr., 288:* 389 (1984).
155. S. A. Cohen, K. P. Benedek, S. Dong, Y. Tapuhi, and B. L. Karger, *Anal. Chem., 56:* 217 (1984).
156. R. H. Ingraham, S. Y. M. Lau, A. K. Taneja, and R. S. Hodges, *J. Chromatogr., 327:* 77 (1985).
157. C. H. Lochmüller and S. S. Saavedra, *Langmuir, 3:* 433 (1987).
158. X. M. Lu, K. Benedek, and B. L. Karger, *J. Chromatogr., 359:* 19 (1986).
159. S. A. Cohen, S. Dong, K. Benedek, and B. L. Karger, in *Affinity Chromatography and Biological Recognition* (I. M. Chaiken, M. Wilchek, and I. Parikh, eds.), Academic Press, New York, p. 479 (1984).
160. W. R. Melander, H. J. Lin, J. Jabobson, and Cs. Horváth, *J. Phys. Chem., 88:* 4527 (1984).
161. W. R. Melander, H. J. Lin, J. Jacobson, and Cs. Horváth, *J. Phys. Chem., 88:* 4536 (1984).
162. D. Pennica, W. E. Holmes, W. J. Kohr, R. N. Harkins, G. A. Vehar, C. A. Ward, W. F. Bennett, E. Yelverton, P. H. Seeburg, H. L. Heynecker, D. V. Goeddel, D. Collen, *Nature, 301:* 214 (1983).
163. G. Pohl, M. Källström, N. Bergsdorf, P. Wallén, and H. Jörnwall, *Biochemistry, 23:* 3701 (1984).
164. K. Benedek, M. B. Seaman, B. R. Hughes, J. K. Swadesh, *J. Chromatogr., 444:* 191 (1988).
165. G. Sofer and C. Mason, *Biotechnology, 5:* 239 (1987).
166. J. Bonnerjea, S. Oh, M. Hoare, and P. Dunnill, *Bio/Technology, 4:* 954 (1986).
167. S. K. Sharma, *Separation Science and Technology, 21:* 701 (1986).
168. B. L. Karger, L. R. Snyder, and Cs. Horváth, *An Introduction to Separation Science,* John Wiley, New York, p. 557 (1973).
169. J. C. Giddings, *HRC CC. J. High Resolut. Chromatogr. Chromatogr. Communic., 10:* 319 (1987).
170. K. Biemann and S. A. Martin, *Mass Spectrometry Reviews, 6:* 1 (1987).
171. M. E. Hemling, *Pharm. Res., 4:* 5 (1987).
172. K. Biemann, *Int. J. Mass Spec. Ion Phys., 45:* 183 (1982).
173. S. A. Carr and G. D. Roberts, *Anal. Biochem., 157:* 396 (1986).
174. S. A. Carr and G. D. Roberts, in *Methods of Protein Sequence Analysis* (K. A. Walsh, ed.), Humana Press, Clifton, New Jersey, p. 423 (1987).
175. Y.-S. Chen, S. V. Brayton, and C. C. Hach, *American Lab., June:* 62 (1988).
176. S. Borman, *Anal. Chem., 59:* 969A (1987).
177. D. B. Wetlaufer, *Adv. Protein Chem., 17:* 323 (1962).
178. J. W. Donovan, *Methods Enzymol., 27:* 497 (1973).
179. D. Freifelder, *Physical Biochemistry—Application to Biochemistry and Molecular Biology,* W. H. Freeman, San Francisco, California (1982).
180. J. Frank, A. Braat, and J. A. Duine, *Anal. Biochem., 162:* 65 (1987).
181. F. Downs and W. Pigman, *Int. J. Pept. Protein Res., 1:* 181 (1969).

182. M. Ikawa and E. E. Snell, *J. Biol. Chem., 236*: 1955 (1961).
183. N. P. Neumann, *Meth. Enzymol., 11*: 487 (1967).
184. C. O. Moorhouse, A. R. Law, and C. Maddix in *Technicon International Congress on Advances in Automated Analysis, Vol. 2*, Mediad, Tarrytown, New York, p. 182 (1977).
185. H. S. Olcott and H. Fraenkel-Conrat, *J. Biol. Chem., 171*: 483 (1947).
186. J. C. Fletcher and A. Robson, *Biochem., J., 87*: 553 (1963).
187. M. W. Rees, *Biochem. J., 40*: 632 (1946).
188. J. K. Swadesh, T. W. Thannhauser, and H. A. Scheraga, *Anal. Biochem., 141*: 397 (1984).
189. A. S. Inglis, *Meth. Enzymol., 91*: 26 (1983).
190. A. S. Inglis and T.-Y. Liu, *J. Biol. Chem., 245*: 112 (1970).
191. C. H. W. Hirs, W. H. Stein, and S. Moore, *J. Biol. Chem., 211*: 941 (1954).
192. S. Moore and W. H. Stein, *J. Biol. Chem., 176*: 367 (1948).
193. S. Udenfriend, S. Stein, P. Böhlen, W. Dairman, W. Leimgruber, and M. Weigele, *Science, 178*: 871 (1972).
194. J. R. Benson and P. E. Hare, *Proc. Natl. Acad. Sci. USA, 72*: 619 (1975).
195. K. S. Lee and D. G. Drescher, *Int. J. Biochem., 9*: 457 (1978).
196. M. Fujiwara, Y. Ishida, N. Nimura, A. Toyama, and T. Kinoshita, *Anal. Biochem., 166*: 72 (1987).
197. S. Einarsson, B. Josefsson, and S. Lagerkvist, *J. Chromatogr., 282*: 609 (1983).
198. N. Kaneda, M. Sato, and K. Yagi, *Anal. Biochem., 127*: 49 (1982).
199. C. DeJong, G. J. Hughes, E. Van Wieringen, and K. J. Wilson, *J. Chromatogr., 241*: 345 (1982).
200. J.-Y. Chang, R. Knecht, and D. G. Braun, *Biochem. J., 199*: 547 (1981).
201. J. D. H. Cooper, G. Ogden, J. McIntosh, and D. C. Turnell, *Anal. Biochem., 142*: 98 (1984).
202. R. L. Heinrikson and S. C. Meredith, *Anal. Biochem., 136*: 65 (1984).
203. G. E. Tarr in *Methods of Protein Microcharacterization* (J. E. Shively, ed.), Humana Press, Clifton, New Jersey, p. 155 (1986).
204. F. S. Esch, *Anal. Biochem., 136*: 39 (1984).
205. N. Takahashi, Y. Takahashi, N. Ishioka, B. S. Blumberg, and F. W. Putnam, *J. Chromatog, 359*: 181 (1986).
206. T. W. Thannhauser, C. A. McWherter, and H. A. Scheraga, *Anal. Biochem., 149*: 322 (1985).
207. S. L. Carney in *Carbohydrate Analysis* (M. F. Chaplin and J. F. Kennedy, eds.), IRL Press, Oxford, p. 97 (1986).
208. E. S. Yeung, L. E. Steenhoek, S. D. Woodruff, and J. C. Kuo, *J. Chromatogr., 52*: 1399 (1980).
209. A. F. Drake and G. D. Jonas, *Chromatog. Anal.*, Feb.: 11 (1989).
210. A. Mannschreck, D. Andert, A. Eiglsperger, E. Gmahl, and H. Buchner, *Chromatografia, 25*: 182 (1988).
211. D. K. Lloyd, D. M. Goodall, and H. Scrivener, *Anal. Chem., 61*: 1238 (1989).
212. B. H. Reitsma and E. S. Yeung, *J. Chromatog, 362*: 353 (1986).
213. J. K. Swadesh and T. A. Perfetti, oral presentation (paper #154) at the 198th American Chemical Society Meeting, Miami, Florida (1989).
214. J. K. Swadesh, *American Lab.* February :72 (1990).
215. P. Cuatresecas in *Drug Development, 2nd Edition* (C. E. Hammer, ed.), CRC Press, Boca Raton, Florida (1990).

Index

Abdruanycin, 86
Absorbance ratios, 124, 132
    ratiogram, 134, 136
Acetaminophen, 74, 82-83, 95, 150, 154
    IR-spectra, 164
Acetylated-β-cyclodextrin chiral phase, 223, 228
Acetylcholine, 79
Acetylcholinesterase, 79
N-Acetyl-D,L-penicillamine, 218
α-Acid glycoprotein chiral phase, 223, 228, 230
ACTH, 261
Adrenergic agents, 215
Alprenolol, 215
Amines, 18
Amino acid chiral phase, 224
Amino acids, 17, 19-20, 28, 273, 290
    chemiluminescence detection of, 33, 35
Amino alcohols, 215-216, 226
Aminoglutehimide, 229
Amperometric transducer (*see also* Electrodes), 67-68
Amphetamine, 215, 226
Ampholytes, 266
Anthracene crystals (see Scintillators)
1-(1-Anthryl)ethylamine, 215
1-(2-Anthryl)ethylamine, 215
Antiasthmas, 89

Antibiotics, 85
Antidepressants, 89
Antiepileptics, 232
Antiinflamatory drugs, 216
Artificial intelligence, 203-204
Arychlorides, 226
L-Asp-L-phe-O-CH$_3$-Cu(II), Zn(II), 218
Aspirin, 150, 154, 174
    IR-spectra, 164
Atrial natriuretic factor, 242
Automated sample preparation
    by column switching, 41-61
    by robotics, 55, 179, 192-196

Barbiturates, 81, 221, 226, 232
Basic chromatographic attributes, 4-5
Benoxaprofen, 226
Benzodiazepin-2-one, 226
Benzodiazepines, 81, 230, 232
Benzothiodiazepines, 232
Betaine, 79
Biogenic amines, 91
Bisantrene, 86
β-Blockers, 89, 215, 220
t-Boc-L-ala-anhydride, 215
t-Boc-L-leu-anhydride, 215
Bovine carbonic anhydrase, 275
Bovine pancreatic trypsin inhibitor, 252

Bovine serum albumin chiral phase, 223, 228, 230
Bushy stunt virus, 260

Caffeine, 150, 153
Captopril, 87–88
Carboquone, 56
Carboxylic acids, 221
Cardiotonics, 89
Catalase, 260
Catecholamines, 29–30, 56, 73, 96, 215
Cefmetazole, 56
Cell culture, 248–249
Cell types, 248
Cellulose carbamate chiral phase, 232
Cellulose triacetate chiral phase, 223, 231
Cellulose tribenzoate chiral phase, 223, 231
Cellulose tribenzyl ether chiral phase, 223, 231
Cellulose tricinnamate chiral phase, 223, 231
Cellulose triphenylcarbonate chiral phase, 223, 231
Cephalosporin, 35, 81
Chemotherapeutics, 85–86
Chiral chromatography, 213–239
Chiral ion pair chromatography, 217–219
  with camphorsulfonate, 220
  with quinine ion, 220
  with tartaric acid ester/ hexafluorophosphate, 220
Chiral mobile phases, 216
Chlomezanon, 232
Chloramphenicol, 85, 126
N-(S)-2-(4-Chlorophenyl)isovaleroyl-(R)-phenylglycine chiral phase, 222
Chlorpheniramine, 229
Chlorthalidone, 229, 232
Choline oxidase, 79
Choline, 79
α-Chymotrypsin, 268
α-Chymotrypsinogen, 262, 264, 268, 270, 275
Ciramadol, 83

Cocaine, 81
Column switching
  accessories, 48
  contribution to band broadening, 44–46
  with non-redundant column, 43–44, 46
  with redundant column, 43–44, 46
Conalbumin, 268, 275
Contour maps, 124, 127
Corticosteroid formulation, 51
Corticosterone, 150, 154
  IR-spectra, 159–161
Cortisone, 150, 154
  IR-spectra, 159–161
Coupled column chromatography, 46
  cation exchange-reversed phase, 50–51, 54
  GPC-reversed phase, 56
  reversed phase-reversed phase, 50–51, 56
γ-Crystallins, 257
Cyclic voltammetry, 71–72
Cyclodextrin
  chiral phase, 219, 223, 226–228
  mobile phase, 221
Cytochrome, 261, 264, 268, 270, 273, 275–276

Dansyl chloride (DNS) - TCPO, 33, 35, 292
Daunorubicin, 86
Derivatization agents, 80
  chiral, 214–216
  for HPLC/FT-IR, 149–150, 165
Dezocine, 83
Diazepam, 15–16
2,2-Dimethoxypropane, 149–150, 165
1-(4-Dimethylamino-1-naphthyl)ethylamine, 215
2,4-Dinitrofluorobenzene, 78
Diode array detector, 123–146
  in dissolution studies, 174, 176
  operation principle, 125
  optical design, 125
  in purity check, 127, 134–138, 289
  in peak identification, 143–145

Index

Diphtheria toxin, 266-267
Dissolution testing, 173-183
   computer control in, 182
   of controlled release dosage forms, 174
      with surfactants, 180
   errors due to
      loss in filter, 179
      sorption on tubings, 180
      wrong timing of sample withdrawal, 177-179
   hardware in
      filter materials, 179
      tubing materials, 180
   in monitoring degradation products, 175
   of multicomponent dosage forms, 175
   sample preparation, 173
      automation in, 179-181
      by HPLC, 175
   of solid dosage forms, 173-176
   of transdermal dosage forms, 176-178
   unit operations, 173
DNA, 260
DNA/RNA fragements, 82-83
DNA transcription, 284
N-DNB-leucine chiral phase, 222
N-DNB-phenylglycine chiral phase, 222
DOPA, 218
Dopamine, 33
Doxorubicin, 75

Edman sequencing, 269, 278, 285, 287, 292
Electrochemical detection (*see also* LCEC)
   of analgesics, 82
   of antiasthmas, 89
   of antibiotics, 85
   of antidepressants, 89
   of aromatic amines, 74
   of biogenic amines, 91
   of β-blockers, 89
   of cardiotonics, 89
   of chemotherapeutics, 85
   of NADH, 75, 79, 93
   parallel mode of operation, 68-69
   of phenolic substances, 73
   of phenothiazines, 75
   of quinones, 75
   response ratios, 69
   series mode of operation, 69-70
   of thiols, 74, 87, 93
Electrodes
   Ag/AgCl, 65
   Au/Hg, 85
   carbon, 67, 75, 82, 85
   mercury, 67, 75
Electron donor/acceptor, 225, 231
β-Emitters, 104
Enantiomeric resolution
   direct, 216-223
   indirect, 214-216
Endoglycosidase, 278-280
Ephedrine, 215, 220, 226
Epidermal growth factor, 242
Equiabsorptive wavelengths, 129
Erythropoietin, 242
Ethotoin, 232

FAB mass spectromethy, 285
Fenoprofen, 226
Fibroblast growth factor, 242
Fludrocortisone acetate, 175
9-Fluorenylmethyl chloroformate, 292
Fluorescamine, 292
Fourier transform infrared (*see* HPLC/FT-IR)

Geiger-Muller counter, 103
Glycoproteins, 263
Glycosidase, 79
Granulocyte colony stimulating factor, 242

Hemoglobin variants, 259
Hexobarbital, 229
High speed HPLC (see also microbore LC)
   band dispersion, 11-15

[High speed HPLC]
  theory, 6–11
Homogeneity, 143, 284
HPLC/FT-IR, 147–169
  by absorbance subtraction techniques, 148
  flow cells, 148
    for transmission technique, 148
    for reflectance technique, 148–149
    post-column reactions in, 149–153
    sensitivity, 165
  by solvent elimination technique, 148–149
    of water, 149, 165
Human growth hormone, 243, 269
Human serum albumin mobile phase, 221
Hydrodynamic voltammetry, 72
α-Hydroxy acids, 218
Hydroxyproline chiral phase, 224
Hyponotic agents, 232

Ibuprofen, 215, 226
Ifosfamide, 232
L-Ileu-Cu(II), 218
Indoprofen, 215
Insulin, 261, 270
Interferon, 243, 269
Interleukin, 243

Ketamine, 56
Ketobemidene, 82
Ketoprofen, 229
Kinematics, 187

Laboratory unit operations (LUO), 189, 201
Lactalbumin, 273, 292
LCEC (*see also* Electrochemical detection), 65–100
  background, 65
  reaction detection schemes, 77–78
  selectivity, 71
Leukotrienes, (see Radiochemical detection)

LH-RH, 261
Ligand exchange chromatography, 216–217
Lysozyme, 275–276

Macrophage colony stimulating factor, 243
Malaria antigens, 243
Mandelic acid, 221
Markovian process, 252
Mass sensitivity (see Microbore LC)
Mephenytoin, 221, 226, 229
Mephobarbital, 229, 232
Methadone, 229
Methaqualone, 232
Methotrexate, 86
4-Methoxylpherine, 226
R-α-Methylbenzyl isothiocyanate, 215
α-Methyldopa, 71, 72, 218
(+) and (–)-α-Methyl-1-napthaleneacetic acid, 215
N-Methyl-L-proline-Cu(II), 218
Metoprolol, 229
Mianserin, 232
Microbore LC (*see also* High speed HPLC), 25–39, 59, 148
  advantages of, 26
  with deuterated solvent, 31
  with information rich detectors, 31
  mass sensitivity of, 26–28
  in multidimension chromatography, 31
  solvent economy of, 29, 31
  stationary phase economy of, 29
  thermal stability of, 31
Microdialysis cannula, 33, 92
Minimum detection level, 26–27
Mitoxantrone, 86
Mono P, 267
Mono Q, 266–267
Morphine, 83
Multidimension chromatography
  (*see* Coupled column chromatography)
Myoglobin, 268, 275–276
Myosin, 260

Index

NADH, 75, 79, 93
Naloxone, 51
Naproxen, 84, 215, 220, 225
1-Napthalenemethylamine, 226
2-Napthylacetylchloride, 226
Napthylalanine chiral phase, 222
2-Napthylaldehyde, 226
β-Napthylchloroformate, 226
(S)-1-(α-Naphthyl)ethylamine chiral phase, 222
1-(α-Naphthyl)ethylaminocarbonylvaline chiral phase, 223
R(-)-1-(Napthyl)ethyl isocyanate, 215
Neochymotrysin, 268
Neurotransmitters (see also Biogenic amines), 32, 69
Niacin, 150, 153, 156
Nicotinamide, 150, 153, 156,
Nifedipine, 87
Nikkomycins, 126-134
    structures, 128
Nimodipene, 229
Ninhydrin, 292
Nisolidipene, 229
Norgestrel, 221, 229
Normetanephrine, 218
Norzimeldine, 124
Nucleic acids, 58
    Nucleobases, 33
    Nucleosides, 33
    Nucleotides, 21, 32, 93

n-Octylamine-Cu(II), 218
Ovalbumin, 264, 270, 275
Oxapadol, 232
Oxprenolol, 215

Paraben preservatives, 54
Peak capacity, 42-43
Peak compression technique, 45-46
Peak cutting technique, 46
    front cut mode, 48, 50
    heart cut mode, 48, 56
Penicillins, 81, 126
L-Phe-Cu(II), 218
Phenacetine, 84

Phenothiazines, 75
Phenprocoumon, 230
Phensuccimide, 229
(R)-(1-phenyl)ethylurea chiral phase, 223
S(-)-1-Phenylethyl isocyanate, 215
Phenyl isothiocyanate, 292
(Phenyl sulfonyl)prolyl chloride, 215
Phosphinothricin (PTC), 33-34
Photochemical/photolytic derivatization, 81
O-Phthalaldehyde (OPA), 16-20, 78-79, 292
Physostigmine, 87-89
Pipecolic acid, 218
Pirkle type chiral phase, 219, 222, 225
Plasminogen activator, 252, 287
cis-Platinum complexes, 85
Poly(2-pyridyldiphenylmethyl methacrylate) chiral phase, 224
Poly(triphenylmethyl methacrylate) chiral phase, 223, 231
Polymeric chiral phase, 232
Post-column enzyme reactions, 79
Praxignantel, 232
Prednisone, 201
Prilocain, 230
Procabazine, 86
Progesterone, 117
L-Proline-Cu(II), 218
Proline chiral phase, 224
Propranolol, 215, 226, 229
Protease activity, 252
Protein
    conformation, 245-246, 270
        effect on separation, 270-272
    denaturation, 250
        effect on separation, 272-275
    disulfide bonds, 251-253
        effect on separation, 275, 279
    heterogeneity, 252-253
    microheterogeneity, 247, 278
    refolding, 249-251
Protein HPLC
    adsorption problems, 255
    in electrostatic interaction mode, 266
    chromatofocusing, 253, 266

Index

[Protein HPLC]
  isoelectric focusing, 266
  gradient elution, 261
  in hydrophobic interaction mode, 266-269
  isocratic elution, 261
  mobile phases, 262-263
  in purity determination, 288
  in reversed phase mode, 269-270, 278
  in size exclusion mode, 263-266
  stationary phases, 256
Pseudoephedrine, 215, 226
Pseudoisomeric graphic, 124
PTC-Ala-Ala, 33-34
PyTechnology, 200

Quinones, 75

Radiochemical detection, 101-122
  in drug metabolism, 118-121
  of laurate, 107, 109, 117-118
  of leukotrienes, 121
  quantitative, 114
  preparative, 113
Radionuclides
  ^{45}Ca, 103
  ^{14}C, 103-104, 110, 112, 117-118
  ^{36}Cl, 103
  ^{3}H, 103-104, 110, 112
  ^{125}I, 103
  ^{131}I, 103
  ^{32}P, 103-104
  ^{35}S, 103
  ^{99}Tc, 103
Retamin, 232
Rhodopsin, 263
Ribonuclease, 253, 264, 278
RNase, 251, 262, 268, 270, 275-276, 278-280
Robot manufacturers
  Cyberfluor, 201-202, 205
  Intellibotics, 206
  Perkin Elmer, 189, 201
  Synchron Instruments, 207
  Zymark robots, 55, 189, 192, 199, 201-202, 205
Robot
  articulating, 201
  in autoinjection, 189-192
  cylindrical, 187-188, 199
  defined, 186
  in extraction, 192-196
  jointed, 187-188
  networking
    Ethernet, 206
    Token Ring, 206
  SCERA, 187
  spherical, 187

Salicylates, 83-84
Scintillation
  liquid, 101, 104-110, 112
    chemical quench of, 104-105
    color quench of, 104-105
    flow cell, 102, 106-107
  solid, 110-111, 112
    flow cell, 110-111
Scintillators, 110-111
  anthracene crystals, 110
  yttrium silicate, 110
Serum albumin, 260
  bovine, 262, 264, 270, 275, 279
SFC/IR, 166
Solvophilic partitioning, 251
Solvophobic partitioning, 251
Spectral suppression, 129
Sterigmatocystin, 57
Stevioside, 56-57
Stoichiometric coefficient (Z), 271-272, 291
Stoichiometric displacement model, 271
Streptomycins, 126
Superoxide dismutase, 243

T4 lysozyme, 269
(R,R) Tartaric acid, 218
  anhydride, 215
Teniposide, 86

Index

N-Tert-butylaminocarbonyl-(S)-valine chiral phase, 223
Test validation, 283
Testosterone, 117
Tetracyclines, 85
2,3,4,6-Tetra-O-acetyl-β-D-glucopyranosyl isothiocyanate, 215–216
α-(2,4,5,7-Tetranitro-9-fluorylideneaminooxy) propionic acid chiral phase, 231
Thalidomide, 232
Theophylline, 143
Thiols, 74, 87, 93
Thyroid hormones, 218
Tissue plasminogen activator, 243
TLC/IR, 147
Tobacco mosaic virus, 260
N-(1R, 3R)-*trans*-Chrysan-themoyl-(R)-phenylglycine chiral phase, 223
2,3,4-Tri-O-acetyl-α-D-arabinopyranosyl isothiocyanate, 215
N-Trifluoroacetyl-S-prolyl chloride, 215
Trinitrobenzene sulfonic acid, 78
Tropic acid, 226
Tropicamide, 226
Trypsin inhibitor, 275
Tumor necrosis factor, 243

Urapidil, 140–142
  structure, 131
Urdamycins, 134–137
  structure, 134
Urinary steroids, 33, 37

L-Valine- Cu(II), 218
Valine chiral phase, 224
Verapamil, 229
Vitamin
  B_2, 57
  K_1, 75, 93
Volipram, 232
von Willebrand factor, 242

Warfarin, 56, 230
Western Blotting, 284, 286

Yttrium silicate (*see* Scintillators)

Z value, 271–272, 291
Zimeldine, 124

Coventry University

£65